*The Millstone Quarries
of Powell County,
Kentucky*

Contributions to Southern Appalachian Studies

1. *Memoirs of Grassy Creek: Growing Up in the Mountains on the Virginia–North Carolina Line.* Zetta Barker Hamby. 1998

2. *The Pond Mountain Chronicle: Self-Portrait of a Southern Appalachian Community.* Leland R. Cooper and Mary Lee Cooper. 1998

3. *Traditional Musicians of the Central Blue Ridge: Old Time, Early Country, Folk and Bluegrass Label Recording Artists, with Discographies.* Marty McGee. 2000

4. *W.R. Trivett, Appalachian Pictureman: Photographs of a Bygone Time.* Ralph E. Lentz, II. 2001

5. *The People of the New River: Oral Histories from the Ashe, Alleghany and Watauga Counties of North Carolina.* Leland R. Cooper and Mary Lee Cooper. 2001

6. *John Fox, Jr., Appalachian Author.* Bill York. 2003

7. *The Thistle and the Brier: Historical Links and Cultural Parallels Between Scotland and Appalachia.* Richard Blaustein. 2003

8. *Tales from Sacred Wind: Coming of Age in Appalachia. The Cratis Williams Chronicles.* Cratis D. Williams. Edited by David Cratis Williams and Patricia D. Beaver. 2003

9. *Willard Gayheart, Appalachian Artist.* Willard Gayheart and Donia S. Eley. 2003

10. *The Forest City Lynching of 1900: Populism, Racism, and White Supremacy in Rutherford County, North Carolina.* J. Timothy Cole. 2003

11. *The Brevard Rosenwald School: Black Education and Community Building in a Southern Appalachian Town, 1920–1966.* Betty Jamerson Reed. 2004

12. *The Bristol Sessions: Writings About the Big Bang of Country Music.* Edited by Charles K. Wolfe and Ted Olson. 2005

13. *Community and Change in the North Carolina Mountains: Oral Histories and Profiles of People from Western Watauga County.* Compiled by Nannie Greene and Catherine Stokes Sheppard. 2006

14. *Ashe County: A History.* Arthur Lloyd Fletcher (1963). New edition, 2006

15. *The New River Controversy.* Thomas J. Schoenbaum (1979). New edition, 2007

16. *The Blue Ridge Parkway by Foot: A Park Ranger's Memoir.* Tim Pegram. 2007

17. *James Still: Critical Essays on the Dean of Appalachian Literature.* Edited by Ted Olson and Kathy H. Olson. 2007

18. *Owsley County, Kentucky, and the Perpetuation of Poverty.* John R. Burch, Jr. 2007

19. *Asheville: A History.* Nan K. Chase. 2007

20. *Southern Appalachian Poetry: An Anthology of Works by 37 Poets.* Edited by Marita Garin. 2008

21. *Ball, Bat and Bitumen: A History of Coalfield Baseball in the Appalachian South.* L.M. Sutter. 2008

22. *The Frontier Nursing Service: America's First Rural Nurse-Midwife Service and School.* Marie Bartlett. 2008

23. *James Still in Interviews, Oral Histories and Memoirs.* Edited by Ted Olson. 2009

24. *The Millstone Quarries of Powell County, Kentucky.* Charles D. Hockensmith. 2009

The Millstone Quarries of Powell County, Kentucky

CHARLES D. HOCKENSMITH
Foreword by NANCY O'MALLEY

CONTRIBUTIONS TO SOUTHERN APPALACHIAN STUDIES, 24

McFarland & Company, Inc., Publishers
Jefferson, North Carolina, and London

Unless otherwise noted, photographs and illustrations are by
Charles D. Hockensmith, Kentucky Heritage Council, Frankfort.

LIBRARY OF CONGRESS CATALOGUING-IN-PUBLICATION DATA

Hockensmith, Charles D.
  The millstone quarries of Powell County, Kentucky /
Charles D. Hockensmith ; foreword by Nancy O'Malley.
    p. cm.   (Contributions to Southern Appalachian studies ; 24)
    Includes bibliographical references and index.

ISBN 978-0-7864-3859-4
softcover : 50# alkaline paper ∞

1. Powell County (Ky.)—Antiquities. 2. Excavations
(Archaeology)—Kentucky—Powell County. 3. Quarries and
quarrying—Kentucky—Powell County—History. 4. Millstones—
Kentucky—Powell County—History. 5. Archaeology and history—
Kentucky—Powell County. 6. Powell County (Ky.)—History, Local.
  7. Powell County (Ky.)—Biography. I. Title. II. Series.
F457.P7H63   2009   976.9'585—dc22   2008051335

British Library cataloguing data are available

©2009 Kentucky Heritage Council. All rights reserved

*No part of this book may be reproduced or transmitted in any form
or by any means, electronic or mechanical, including photocopying
or recording, or by any information storage and retrieval system,
without permission in writing from the publisher.*

On the cover: Millstone #11 at the Pilot Knob Millstone Quarry,
Powell County, Kentucky, 1988 (photograph by the author);
inset: Old Grist Mill II ©2009 Shutterstock.

Manufactured in the United States of America

*McFarland & Company, Inc., Publishers
Box 611, Jefferson, North Carolina 28640
www.mcfarlandpub.com*

Table of Contents

Acknowledgments ix
Foreword by Nancy O'Malley 1
Preface 3
Introduction 7

1. AN ARCHIVAL OVERVIEW OF THE KENTUCKY MILLSTONE
   QUARRYING INDUSTRY   11

   Franklin County   11
   Letcher County   12
   Logan County   13
   Madison County   13
   Marshall County   14
   McCreary County   14
   Powell County   14
   Rockcastle County   15
   Whitley County   17
   Woodford County   17
   General Comments   18

2. THE ENVIRONMENTAL SETTING OF THE QUARRIES   19
   Geology   19
   Soils   20
   Flora   20

3. DOCUMENTING THE QUARRIES: FIELD METHODS, FIELD LIMITATIONS,
   AND ARCHIVAL RESEARCH   22
   Field Methods   22
   Field Limitations   23
   Archival Research   24

4. AN HISTORICAL OVERVIEW OF THE POWELL COUNTY QUARRIES   26

5. FAMILIES ASSOCIATED WITH THE POWELL COUNTY QUARRIES   30

   Adams Family   30
   Berry Family   33
   Daniel Family   35
   DeWitt Family   36
   Golf Family   37
   Hanks Family   38
   Hedger Family   38
   Johnson Family   38
   Nelson Family   39
   Pigg Family   39
   Risk Family   40
   Ross Family   41
   Smith Family   41
   Spry Family   41
   Stewart Family   42
   Summers Family   42
   Treadway Family   43
   Ware Family   45
   West Family   45
   Stone Cutters   45

6. ARCHAEOLOGICAL INVESTIGATIONS AT THE POWELL COUNTY QUARRIES   47
    McGuire Quarry   47
    Baker Quarry   63
    Toler Quarry   76
    Ware Quarry   84
    Ewen Quarry   98
    Pilot Knob Quarry   104

7. COMPARISONS AMONG THE POWELL COUNTY QUARRIES   112
    Millstones   112
    Leveling Crosses   115
    Cutting Eyes   117
    Drill Holes on Millstones   118
    Drilled Boulders   119
    Shaping Debris   120
    Quarry Excavations   122
    Artifacts   123
    Tool Marks on Boulders and
       Millstones   127
    Reasons for Millstone Rejection   132

8. COMPARISONS BETWEEN THE POWELL COUNTY QUARRIES
    AND OTHER QUARRIES   134
    Leveling Crosses   134
    Reasons for Millstone Rejection   135
    Quarry Excavations   136
    Tools   136

9. MANUFACTURING SEQUENCES FOR MILLSTONES   138
    Techniques in Europe   138
    Techniques in the United States   139

10. TRANSPORTATION METHODS AND ROUTES FOR POWELL COUNTY
    MILLSTONES   148
    Road Transport   148
    River Transport   149

11. MARKETS FOR POWELL COUNTY MILLSTONES   150

12. THE COMPETITION: IMPORTED MILLSTONES IN KENTUCKY   152

13. KENTUCKY MILLSTONE VALUES   157

14. CONCLUSIONS   159

Glossary   163
Appendices
    A: Form for Documenting Millstones   167
    B: Form for Documenting Boulders and Drill Holes   170
    C: 1804 Lawsuit in Fayette County, Higbee v. Hanks   172
    D: 1810 Lawsuit in Clark County, Wilkerson v. Adams   173
    E: 1823 Lawsuit in Clark County, Johnson v. Adams   175
    F: 1826 Lawsuit in Clark County, Summers v. Adams   177
    G: Millstones Potentially Associated with the Powell County Quarries   178
Bibliography   181
Index   197

Acknowledgments

This book is the culmination of four seasons of fieldwork and many years of part time archival research. During this period, many people have contributed to this study in numerous ways. Without their assistance, this study would have been far less comprehensive. These numerous academic debts are acknowledged below. Because of the length of time that has passed since the beginning of this research, these acknowledgments are organized roughly chronologically.

First and foremost, I wish to express my appreciation to the Kentucky Heritage Council, whose financial support made this study possible. I was given staff time to conduct archival research and write, in connection with my research on the Kentucky millstone quarries, in my capacity as a staff archaeologist. Since its creation in 1966, the Kentucky Heritage Council has taken the lead in preserving and protecting Kentucky's cultural resources. To accomplish its legislative charge, the Heritage Council maintains three program areas: Site Development, Site Identification, and Site Protection and Archaeology (under which this work was accomplished). The Site Protection and Archaeology Program staff works with a variety of federal and state agencies, local governments, and individuals to assist them in complying with Section 106 of the National Historic Preservation Act of 1966 and to ensure that potential impacts to significant cultural resources are adequately addressed prior to the implementation of federally funded or licensed projects. The Site Protection and Archaeology Program is also responsible for administering the Heritage Council's archaeological programs, organizing archaeological conferences, editing and publishing volumes of selected conference papers, and disseminating educational materials. On occasion, the Site Protection and Archaeology Program staff undertakes field and research projects, such as emergency data recovery at threatened sites. The agency can be contacted at the following address: Kentucky Heritage Council; 300 Washington Street; Frankfort, Kentucky 40601 or by phone at (502) 564–7005.

Past and present staff members of the Kentucky Heritage Council provided assistance with this study. David L. Morgan, former Director of the Kentucky Heritage Council, and Thomas N. Sanders, former Site Protection Program Manager, provided the encouragement and resources necessary to undertake and complete this project. In 2005, David Pollack encouraged the completion of this book. Donna M. Neary, former Director of the Kentucky Heritage Council, graciously co-signed the book contract and encouraged me. I am very grateful to David, Tom, David, Donna, and the Kentucky Heritage Council for their faithful support of this project through the years.

During the four field seasons, several individuals assisted me with fieldwork. These include Johnny Kimbrell (1987–1989), Larry Meadows (1987–1990), Verlin Hasley (1987, 1990), Rhondle Lee (1987), Leif Meadows (1989–1990), Wayne Webb (1987), Roland Herzel (1988), Joe Briggs (1988), and Dale Gafney (1989). A special thanks is due to Johnny Kimbrell who volunteered more days than anyone else. Unfortunately, Johnny passed away several years before this book was completed. We are also indebted to Wayne Webb who shared the results of his metal detecting at the Baker and Ware quarries and graciously loaned the Red River Historical Society the metal artifacts that he found. Several individuals assisted Larry Meadows in searching for new quarries. These include Verlin Hasley, Johnny Faulkner, Wayne Webb, Dale Gafney, Leif Meadows, Ed Drake, and David Shearer.

The archival research benefited from the efforts of several people. Larry Meadows examined deeds and many other early records in the courthouses of Powell, Clark, and Montgomery counties. He also located two early lawsuits that mentioned the quarries. Tom Martin found an early document in the Montgomery County records which he shared. Mr. Bill Adams (Eastern Kentucky University, 1981) provided copies of the documents relating to his ancestor Spencer Adams who was one of the millstone makers in Powell County. I searched the various records (family, archaeological, geological, census, tax, county histories, etc.) housed at various University of Kentucky libraries, the Kentucky Historical Society library, and other libraries. I also searched for additional early lawsuits at the Kentucky Department of Libraries & Archives.

Larry Meadows discovered the millstones quarries in Powell County and brought them to my attention. He had a tremendous desire to see these quarries documented by a professional archaeologist. Larry provided the encouragement and handled local logistics to make the project happen. He is given co-authorship of several chapters of this book in recognition of his many important contributions to this study. Over the course of the study, Larry and I became close friends. Many pleasant conversations about millstone quarrying occurred in the evenings following fieldwork. Without Larry's foresight and determination, this study would not have occurred.

Several colleagues in Kentucky provided assistance. Mr. Ron Bryant, formerly with the Kentucky Historical Society, provided very helpful guidance concerning historical journals and documents available in their library. Ms. Pam Lyons, Kentucky Department of Libraries and Archives, was extremely helpful in obtaining mill and millstone publications through interlibrary loan. Ms. Gayle Alvis, Kentucky Department of Libraries and Archives, made me aware of a great website for searching libraries and assisted in obtaining information. Mr. Cecil Ison, formerly with the U.S. Forest Service, Winchester, Kentucky, shared data from historical documents that he encountered. Dr. R. Berle Clay, Cultural Resource Analysis, Inc., shared information on millstones from Clay family records relating to mills and other information. Mr. Donald B. Ball, formerly with the U.S. Army Corps of Engineers, Louisville District, shared links to new websites that yielded a wealth of previously undiscovered information on American millstone quarries. Mr. Charles M. Niquette, Cultural Resource Analysis, Inc., brought to my attention an 1804 lawsuit filed in Fayette County that mentioned a Powell County millstone quarry. The late Tom Sussenbach also shared information on two possible millstone quarry locations in southeast Kentucky.

Many people provided information on millstones and millstone quarrying in the early stages of this research. These include Dr. Fred E. Coy, Jr. (Louisville, Kentucky), the late Tom Fuller (Louisville, Kentucky), David Rotenizer (Archaeological Society of Virginia, Blacksburg, Virginia), Dale Collins (Pembroke, Virginia), Claude V. Jackson (Tidewater Atlantic Research, Washington, North Carolina), Steve Rogers (Tennessee Historical Commission, Nashville, Tennessee), and geologists Benjamin Gildersleeve (Bowling Green, Kentucky), John Rice Irwin (Museum of Appalachia, Norris, Tennessee), Calvert McIlhany (Bristol, Virginia), Robert G. Schmidt (Arlington, Virginia), and Phil Gettel (Floyd, Virginia).

In an effort to locate missed or unpublished information on millstone quarries in other states, I sent letters to all State Historic Preservation Officers (SHPOs) in the United States during January 1991. Twenty-eight states responded to the information request. Also, I sent letters to additional persons recommended by the SHPOs. From this mailing I received several responses that provided information on conglomerate millstone quarries. Among the individuals responding were Julia S. Stokes (New York State Office of Parks, Recreation and Historic Preservation, Albany), Jerry A. Clouse (Pennsylvania Historical and Museum Commission, Harrisburg), and Bruce B. Brown (Greencastle, Pennsylvania). Special thanks are due to John McGrain, who shared copies of his fact cards on millstones that included many early sources.

Several retired Kentuckians who reside in other states were extremely helpful in providing clues on a millstone quarry in their native Letcher County. They were interviewed by phone in 1992. Mr. Clarence Halcomb (Hamilton, Ohio) provided information about the quarry and its location. Other Letcher county natives who provided information are Mr. Andy Frazier (Madison, Indiana), Mrs. Gladys Hogg (Florida), and Mr. Dover Cornett (Florida).

The current and past owners of the Powell County millstone quarries deserve very special thanks for allowing us to document the quarry remains on their property. These include the late Mr. Kelly McGuire (McGuire Quarry), Mr. James Baker and Mr. Ronnie Baker (Baker and Ware Quarries), Mr. Trigger Toler (Toler Quarry), the late Mrs. A. L. Ewen (Ewen Quarry), and Mr. Richard Hannan, formerly with the Kentucky Nature Preserves Commission (Pilot Knob Quarry). In the years following the fieldwork, all but one of the quarries changed ownership. The Baker and Ware Quarries were acquired by the Kentucky Nature Preserves Commission in 1998 and will be preserved for future generations. Likewise, the McGuire Millstone Quarry was acquired by Powell County in 2002 to ensure its future protection (Meadows 2002). The Red River Historical Society and Museum worked hard to ensure that the Baker, McGuire, and Ware millstone quarries were preserved. The Ewen and Toler millstone quarries were also sold and have new owners. Powell County is currently in the process of purchasing the property containing the Ewen Quarry. It is hoped that the other millstone quarries can be brought into public ownership in the future.

During May 1990, Fred E. Coy, Jr., and I had a unique opportunity to collect significant information on millstone quarrying techniques. Over a two-day period, we were able to interview the last two living millstone makers in Virginia — Mr. Robert Houston Surface and Mr. W. C. Saville — and visit the Brush Mountain Millstone Quarry near Blacksburg where these men were once employed. We are grateful to these gentlemen for sharing their tremendous knowledge concerning the manufacture of millstones. Since no detailed published accounts of millstone making in the United States have come to light, the knowledge possessed by Mr. Surface and Mr. Saville is extremely significant in understanding how conglomerate millstones were manufactured. Unfortunately, Mr. Surface passed away on May 3, 1998, before the completion of this book. Mr. Saville lived a few years longer, passing away on April 11, 2003. Although these men are now gone, much of their knowledge of millstone making lives on through their interviews (see Hockensmith and Coy 1999; Hockensmith and Price 1999).

In another attempt to obtain information on millstone quarries in other states, in 1996 I sent letters to the geological surveys in states that were known commercial producers of millstones. These states all provided some information that supplemented what we had already collected. Special thanks are due to Lewis S. Dean (geologist with the Geological Survey of Alabama, Tuscaloosa), William Kelly (State Geologist with the Geological Survey at the New York State Museum, Albany), Robert C. Smith, II (economic geochemist with the Bureau of

Topographic and Geologic Survey, Pennsylvania Department of Conservation and Natural Resources, Harrisburg), Charles H. Gardner (Director and State Geologist, State of North Carolina Department of Environment, Health, and Natural Resources, Raleigh), and Palmer C. Sweet (Commonwealth of Virginia, Department of Mines, Minerals, and Energy, Charlottesville). Also, very special thanks to Garland R. Dever (retired geologist with the Kentucky Geological Survey, Lexington) for providing addresses of other geological surveys and assisting in a variety of ways through the years.

Museums and historical societies in New York and Pennsylvania were very helpful in obtaining information. Ms. Patricia Christian with the Ellenville Public Library and Museum in Ellenville, New York, provided information on the millstone industry in Ulster County. The late Ms. Eleanor S. Rosakranase and Amanda C. Jones of the Ulster County Historical Society in Kingston, New York, were helpful. Ms. Rosakranase graciously shared copies of *The Accordian,* which included information on the Lawrence brothers and the Ulster County millstone industry. Ms. Cynthia Marquet with the Historical Society of the Cocalico Valley in Ephrata, Pennsylvania, shared copies of photographs of millstones from the Turkey Hill millstone quarry in Lancaster County. Also, Ms. Ruth Baer Gembe, Alexander Hamilton Memorial Free Library, in Waynesboro, Pennsylvania, provided copies of reference materials on millstones and copies of photographs of a millstone collection housed at the library.

In April of 1998, I took a week of vacation time and traveled to New York and Pennsylvania. I was accompanied by my good friend and fellow scholar Dr. Fred E. Coy, Jr., of Louisville, Kentucky. We visited some of the famous Ulster County, New York, millstone quarries near the community of Accord. During the trip, we also interviewed Vincent and Wally Lawrence, two brothers over 80 years of age, whose father and uncles were millstone makers. The Lawrence brothers shared their memories of millstone making during their youth. Unfortunately, the Lawrence brothers did not live to see the interviews published. We were also fortunate to meet Lewis Waruch (whose mother's family included millstone makers), who had been interested in the millstone quarries near Accord for many years. Lewis shared his knowledge freely, took us to several millstone quarries, and showed us his tremendous collection of millstone making tools. We wish to express our gratitude to these men for being willing to share their knowledge about the manufacture of millstones. Dr. Coy videotaped the interviews and scenes at the millstone quarries. The Society for the Preservation of Old Mills funded the transcription of the New York interviews. Janet Gates of Frankfort did an excellent job in transcribing these interviews. The interviews will published in book form (Hockensmith 2008b).

During 1998, Steve Knox, Red River Historical Society, used a metal detector at the Ware and Baker millstone quarries under the supervision of Larry Meadows. Leif Meadows utilized his archaeological training to lay out a grid system that was used during the metal detecting. This grid system allowed the investigators to maintain the provenance of the artifacts recovered. Those artifacts are curated at the Red River Historical Society Museum in Clay City. Twenty-four of the artifacts discovered by Steve Knox and Wayne Webb are described in this book.

A special thank you is owed to the Hagley Museum and Library in Wilmington, Delaware. The Hagley Library kindly provided a copy of Hehnly & Wike's 1880 broadside advertisement announcing "The Newly Discovered Turkey Hill Stones Are the Best Mill Stones in the Country, and in Consequence we have Made a reduced Price List of Cocalico Mill Stones" in Durlach, Pennsylvania. We appreciate their willingness to permit us to use the important information in this one-page advertisement. Also, Mr. Richard Brown of Louisville, Kentucky, found and shared information on the Shacklett family, who made millstones at Laurel Hill in Fayette County, Pennsylvania. This information filled an important void in our records concerning this family.

As this volume approached completion, several people provided assistance. Dr. Harry Enoch of Clark County, Kentucky, kindly provided verbal commentary on individuals mentioned in early Clark County lawsuits. Harry also shared references for ads for Red River millstones dating to 1799 and 1818. The Kentucky Department of Libraries & Archives staff assisted me in locating early Clark County lawsuits which I searched for data concerning the millstone quarries. Staff member Jim Prichard was extremely helpful in obtaining copies of early lawsuits. Ms. Pam Lyons, the interlibrary loan department, Kentucky Department of Libraries & Archives, was very helpful in obtaining out of print books. Mr. Gary Adams of Winchester, Kentucky, graciously granted permission for the use of his drawings previously published in a brochure (Hockensmith 1994a). The Filson Historical Society in Louisville, Kentucky, gave permission for the quoting of an 1824 contract between Charles Colyer and Sidney Payne Clay from the Sidney Payne Clay Papers in their Special Collections. Special thanks are due to Jim Holmberg, Curator of Special Collections, and Jacob Lee, Manuscript Cataloger, of the Filson Historical Society staff for their assistance. Mr. Tim Tingle, Kentucky Department of Libraries & Archives, Frankfort, graciously provided a letter concerning the use of four early lawsuits from their collections. Mr. John B. Skiba, Manager of the Office of Cartography and Publications, New York State Museum, Albany, New York, give permission for quoting from New York State Museum bulletins and other geological publications.

Several organizations allowed me use information previously in my articles and books. Ms. Esther Middlewood, Editor of *Old Mill News,* allowed me to use information from my articles published in *Old Mill News* and my books published by the Society for the Preservation of Old Mills. Dr. Kit Wesler permitted me to use information from my articles in *Ohio Valley Historical Archaeology* and the book jointly published by the Symposium on Ohio Valley Urban and Historic Archaeology. Mr. Steve Spring, Editor of *The Mill Monitor*, allowed me to use data from my article that appeared in *The Mill Monitor* published by the International Molinology Society of America. Dr. Fritz Mangartz of the Römisch-Germanisches Zentralmuseum, Mainz, Germany, allowed me to use information from my article in their book. Mr. Bernard Cesari, Manager of Éditions Ibis Press, Paris, France, graciously permitted me to use information from my paper contained in a book published by Éditions Ibis Press.

A number of organizations kindly allowed me to use quotes from their publications and sent me letters. Ms. Alice Cross, Editor of *The Accordian* published by the Friends of Historic Rochester, Inc., in Accord, New York, allowed me to use quotes. Ms. Esther Middlewood, Editor of *Old Mill News,* granted me permission to use information published in *Old Mill News* and Society for the Preservation of Old Mills publications. Dr. Edward E. Erb, State Geologist and Division Director of Division of Mineral Resources, Charlottesville, Virginia, granted permission to quote from Campbell (1925). Mr. P. Patrick Leahy, the American Geological Institute, authorized quotes from Gary, McAfee and Wolf (1974). Ms. Tamara G. Miller with the Historical Society of Pennsylvania, Lancaster, Pennsylvania, allowed the use of quotes from *The Pennsylvania Magazine of History and Biography*. Dr. Thomas R. Ryan, President & CEO of the Lancaster County Historical Society, permitted me to use quotes from Flory (1951a, 1951b). Mr John R. Keith, Chief, Eastern Region Publications, with the U. S. Geological Survey granted permission to use quotes from the *Mineral Resources of the United States* and the later series the *Minerals Yearbook*. Ms. Michelle L. Mullenax-McKinnie, Vice President of Publishing, McClain Printing Company of Parsons, West Virginia, gave permission to quote from Maxwell (1968).

Two organizations granted permission to use maps that they produced. First, the Kentucky Commerce Cabinet allowed us to reproduce the "Kentucky Base Map Series B-5," compiled and distributed by Kentucky Department of Commerce, Frankfort, Kentucky, in 1964 (Figure 1). We appreciate the assistance of Mr. William R. Dexter, Executive Director of the Office of Legal Affairs, with the Kentucky Commerce Cabinet, Frankfort. Second, the Kentucky Geological Survey permitted us to copy information from their "Geologic Map of Powell County, Kentucky" by Raymond Miller and Guy H. Briggs, Jr. (1929), for our Figure 3. The assistance of Michael J. Lynch, Communications and Technology Transfer, Kentucky Geological Survey, Lexington, Kentucky, is greatly appreciated.

I would also like to thank other people who contributed to the production of this book through the many years of its preparation. Archaeologist and mill scholar Nancy O'Malley (Museum of Anthropology, University of Kentucky) graciously read the entire manuscript and offered many helpful suggestions. Nancy also consented to writing a foreword to this book, for which I am indebted to her. Sarah Miller (formerly with the Kentucky Heritage Council and Kentucky Archaeological Survey) looked at a portion of the manuscript and provided encouragement. Mark Dennen, Kentucky Heritage Council architect, graciously created some circle templates for some figures. Former Kentucky Heritage Council staff members Tracy A. Polsgrove and Rose Murphy assisted with typing some early draft sections of this book. Yvonne Sherrick, Kentucky Heritage Council, also provided assistance during the final preparation of the book. Special thanks are due to the staff of the Photography Section in the Finance & Administration Cabinet of State Government. Gary Robinson (former supervisor, now retired), David Bryan, and Steve Mitchell (now retired) did a great job in printing several batches of photographs of the Powell County millstone quarries. It was always a great pleasure working with these gentlemen in the Photography Section. Esther Middlewood (editor of *Old Mill News*) and Kevin Johnson (president of SPOOM) have graciously referred inquiries about millstones to me. As a result of these inquires, I have been able to correspond with several individuals who have shared important information on the millstone industry that I would not have found otherwise. I thank the Red River Historical Society Museum for loaning 24 of the best-preserved metal artifacts from the Powell County millstone quarries to be measured and photographed.

Finally, I want to express my gratitude to my wife, Susie, who accompanied me to millstone conferences, visited millstone quarries with me, and provided moral support.

Charles D. Hockensmith • Frankfort, Kentucky • Spring 2009

Foreword by Nancy O'Malley

Mills were once such a ubiquitous part of the American landscape that no one was unaware of or unaffected by them. Bread served at every meal, boards used to construct buildings, cotton spun into thread, and many more ordinary commodities necessary for comfortable living — all these things were made possible by mills.

Millstones were an integral part of the mill operation, as necessary as the mill dam or the mill building itself. They were the workhorse component of the mill, and the success of the mill in producing flour, meal, or other products depended on their functioning properly. Millstone manufacture required special skills and hard manual labor, and it carried serious health risks for millstone workers. Only certain types of rock formations were suitable for millstones, and these rocks were often located in remote places that were difficult to reach. Once millstones were set in place and put to work, they required regular maintenance to operate efficiently and effectively. They wore out and had to be replaced; they constituted a significant proportion of the cost of doing business.

Old mill enthusiasts are a unique breed. They look at a creek and see not a scenic gravel shoal but the remnants of a mill dam. They surf the Internet for historic postcards showing old mills. They form organizations solely dedicated to the preservation of old mills. The more scholarly among them spend countless hours on research in old newspapers, deeds, court order books, legal case files, and legislative acts. Charles D. Hockensmith falls under the category of a mill scholar extraordinaire. He has compiled in this book an incredibly comprehensive and exhaustive reference work on millstones. Those of us who are privileged to know Charles know that he has the steadfast commitment and patience of a true scholar who leaves no stone unturned (although he admits to leaving some millstones unturned!) to gather as much information about his subject as he can. He not only spends many hours among archival sources, but he also uses the specialized methods of an archaeologist to identify and record sites associated with millstones. Charles has a predilection for studying topics that other scholars have overlooked or neglected, and his millstone research is just one example. In that research, he provides important reference information for the rest of us, and makes us aware of aspects of historic material culture that we don't usually consider but should as we conduct our own mill research.

In his millstone research, Charles was aided immeasurably by Larry Meadows. Larry is one of those rare individuals who truly can be called the quintessential Renaissance man. Interested in just about anything and everything cultural, he was the one who got Charles interested in millstones and millstone quarries and he has, in his own way, made important contributions to this volume. He will tell you he is not a scholar, not a writer, not an expert — but when Larry Meadows gets interested in a topic, he not only sets out to find out more about it with the single-mindedness of a bloodhound, but he has an uncanny knack for getting other people involved. The two made a formidable duo, each complementing one another's special talents and capabilities.

The resulting book on millstones is a great reference work that brings together years of dedicated scholarship between two covers. It is a significant addition to the mill literature and deserves a place in every mill enthusiast's library as well as in the reference sections of libraries. Charles and Larry both are to be commended for the thoroughness of their coverage and the valuable quantitative and qualitative data presented.

Nancy O'Malley • Assistant Director
William S. Webb Museum of Anthropology • University of Kentucky • Lexington

Preface

During the spring of 1987, the Kentucky Heritage Council and the Red River Historical Society initiated a joint research project to document the millstone quarries of Powell County, Kentucky (Figure 1). The seeds for this research were sown on September 14, 1983, when Larry Meadows and Rhondle Lee dropped by my Frankfort office and shared information about the millstone quarries and other local industries in Powell County (Hockensmith 1983). These quarries sounded very interesting, and four years later I visited two of them. I was so impressed with these quarries that I immediately made plans to begin fieldwork. The first field season occurred during March and April of 1987. During 1988, 1989, and 1990, the fieldwork continued as additional quarries were discovered.

The initial goals of this project were two: first, to collect sufficient information about the Powell County quarries to nominate them to the National Register of Historic Places; and second, to publish a detailed monograph containing the archeological and archival data on this important but little known historic industry. This book fulfills the second goal.

Information contained in this book was gradually compiled over a period of several years. My original, optimistic time frame was redrawn again and again as job responsibilities and other writing obligations took priority. As it turned out, however, the delay made for a bigger and better book. When this research was initiated, I knew very little about the millstone industry. As the years passed I discovered many new publications and made numerous contacts with scholars in the United States and several other countries. I wrote articles and conference papers dealing with the millstone industry, and the research for those pieces greatly increased my knowledge. Also, in recent years many new websites have made available literature (both old and new) that I would have never found without the Internet. Hence the delay in completing this book turned out to be a tremendous blessing, allowing for the creation of a much more comprehensive publication that will be of greater utility to other researchers.

This book deals with the Kentucky millstone quarries and how they compare to other quarries. In the first chapter, archival and other information is presented for millstone quarries in Franklin, Letcher, Logan, McCreary, Madison, Marshall, Powell, Rockcastle, Whitley, and Woodford counties. The environmental setting of the millstone quarries is addressed in a chapter that discusses geology, soils, and flora. Next, a chapter discusses field methods, field limitations, and archival research of the project. An historical overview of the Powell County millstone quarries is presented in the fourth chapter. Utilizing available records, the fifth chapter discusses the families associated with the quarries and other individuals mentioned in the deeds. The sixth chapter describes the millstone quarries and the archaeological remains associated with them. The seventh chapter summarizes and compares data on the six Powell County quarries, and the eighth chapter compares some of that data to information about quarries elsewhere. Next, a chapter discusses the manufacturing sequences for mill-

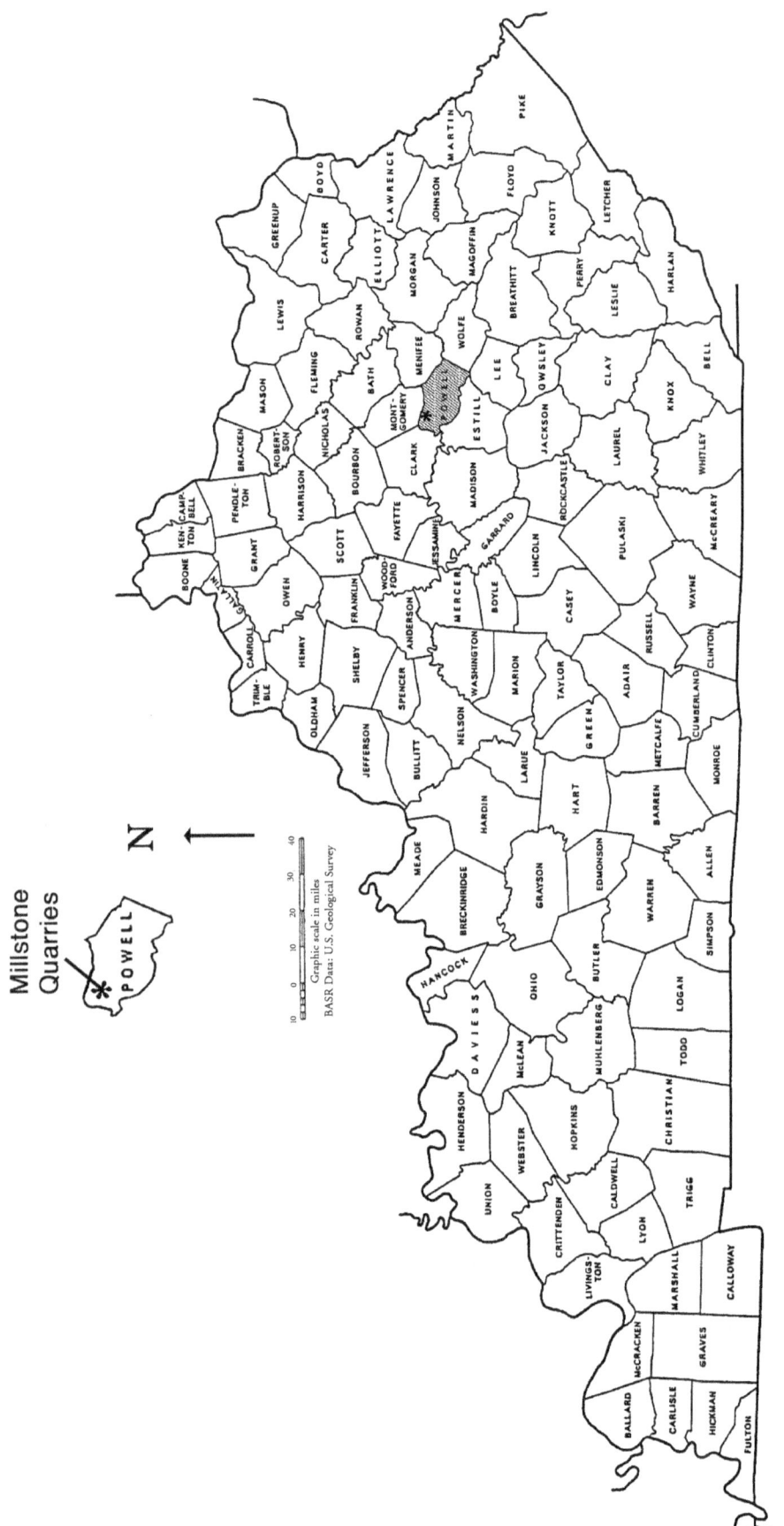

Figure 1. Map of Kentucky, with added detail of Powell County, showing the location of the Powell County millstone quarries. Base Map Series B-5, compiled and distributed by Kentucky Department of Commerce, Frankfort, Kentucky in 1964 (reproduced with the permission of the Kentucky Commerce Cabinet, Frankfort).

stones, comparing methods used in England, in the United States in general, and in Powell County in particular. Four brief chapters then provide information on transportation routes, millstone markets, imported millstones coming into Kentucky, and millstone values in Kentucky. The concluding chapter provides brief final remarks on this study. A glossary is provided to help the reader with some specialized and technical terms. The appendices at the end of the book include modified copies of the recording forms used for millstones and boulders, selected text from lawsuits involving the millstone quarries, and information on millstones that may be associated with the quarries. The book concludes with bibliography and index.

Introduction

The millstone quarries of Powell County, Kentucky, are representative of a poorly known but very important type of historic archaeological site. At one time, thousands of grist mills dotted the American landscape. Each mill required one or more sets of millstones to grind grains. The literature on mills is quite extensive, but little has been written about the stones that were such an essential component of the grinding process.

Powell County was an important millstone producer for Kentucky but not for the nation as a whole. The best known millstone quarries were located in New York, Virginia, and Pennsylvania, where the industry — active in America since the mid–1700s — survived longer than in other parts of the country, into the twentieth century. The Kentucky quarries were probably closed before the U.S. government began collecting information on the millstone industry.

For a more comprehensive overview of the millstone industry, interested readers should consult my book *The Millstone Industry: A Summary of Research on Quarries and Producers in the United States, Europe and Elsewhere* (McFarland, 2009).

Before proceeding further, it is necessary to define two key terms used in this study: millstone and quarry. According to *Webster's New Twentieth Century Dictionary of the English Language* (McKechnie 1978:1143), the term millstone refers to "either of a pair of large, flat, round stones used for grinding grain or other substances." A quarry is defined as a "place where stone or slate is excavated, as by cutting or blasting, for building purposes, etc.: it is usually open to the light, and in this respect differs from a mine" (McKechnie 1978: 1475). This book deals with those locations where conglomerate stone was quarried and subsequently manufactured into millstones. These sites do not have deep pit-type excavations commonly associated with modern quarries. Instead, some of the Powell County quarries contain a series of shallow pits; in other quarries, scattered surface boulders were shaped into millstones.

Grist mills required at least two millstones to grind grains (Figure 2). Larger mills often employed several sets of millstones to increase their production. Typically, millstones were used in pairs, with one stone running above the other stone. The lower millstone was called the bedstone and it remained stationary. The upper millstone was known as the runner and it rotated. Grooves or furrows were cut into the grinding surface of each millstone to facilitate grinding. Different patterns of furrows were used at various times and for varying grinding tasks. Both millstones had to be balanced. The distance between the stones was carefully regulated to prevent them from touching but keep them running close enough together to grind. The upper millstone was attached to the power source through iron hardware (rynd and spindle) and turned by a shaft called a damsel. A wooden housing covered the millstones, and the flour or meal flowed out a spout where it was collected.

Millstones vary according to their design, grinding surface, raw material, and function. Millstones made from a single piece of rock are known as monolithic stones. Composite millstones are built from several small shaped stones which are cemented together and bound with iron bands. Millstones that operated horizontally were called face-grinders while millstones that ran vertically on their edges were called edge-runners or crushers (Tucker 1977:1). Pairs of edge runners attached to the same axle were called chasers. Phalen (1908:610) stated that "chasers are larger than regular millstones. They are used for heavier work, as in grinding quartz, feldspar, barytes, etc., and as already mentioned, run on edge. They were made with a diameter as short as 24 inches, they are usually turned out with diameters ranging from 50 inches to 84 inches and with thicknesses as great as 22 inches." Several types of stone were used for manufacturing millstones. These included conglomerate, fresh water quartz, granite, flint, sandstone, gneiss, quartzite, basalt, and occasionally other types of rock. Small hand turned millstones were called querns and were brought to Kentucky by early pioneers.

Most people think that millstones were used only for grinding corn, wheat, and other grains, but in fact they had many other applications. Sass (1984:10–32, 55–56) has noted that special millstones were used for cleaning clover seeds, shelling oats, hulling buckwheat, pearling barley, processing split peas, chaffing wheat, regrinding middlings, making apple cider, and grinding phosphate rock. Another use was olive oil presses (Kardulias and Runnels 1995:110; Runnels 1981:225–226). They were also used in the chocolate industry, cork mills, dye mills, flint grinding mills, hemp mills, paint and color mills, plaster of Paris and gypsum grinding mills, and tanbark mills (Sass 1984:33–60). Webb (1935:217–218) also notes their use in flax mills, snuff mills, and gunpowder mills. Other applications included grinding bone (Parker 1896:927; Bost 2002:32), mica (Pratt 1901:793; Pratt 1902:793), charcoal (Williams 1885:712), cement (Day 1892:456; Phalen 1908:609; Pratt 1902:794; Pratt 1904a:879; Pratt 1905:1004), barytes (Pratt 1902:794; Pratt 1904a: 879), drugs (Pratt 1905:1004), mustard (Pratt 1905:1004), glucose (Pratt 1905:1004), spices (Pratt 1905:1004), fertilizers (Pratt 1902:794; Pratt 1905:1004), plaster (Pratt 1905: 1004), paste (Pratt 1905:1004), feldspar (Katz 1917:67), quartz (Katz 1917:67), and talc (Phalen 1908:609). A localized task for millstones was the hulling of rice on the southeast coast of America (Judd 1999, 2000).

The millstones quarries described in this book are located in northwest Powell County, Kentucky (Figure 3). Powell County is situated along the east central boundary of the Knobs Region, a semicircular band of high knobs or hills that surround the Bluegrass Region. Six quarry sites have been documented in this area. Three of these quarries are located at Rotten Point, two at Kit Point, and one at Pilot Knob.

The millstones manufactured at the Powell County quarries are all monolithic (one piece) and are made from a Pennsylvanian age (320–284 mybp) conglomerate. This conglomeratic sandstone typically ranges from a light gray to light tan and contains numerous rounded quartz pebbles. It outcrops near the crest of two knobs and also occurs in boulder form on the steep slopes and narrow ridges forming

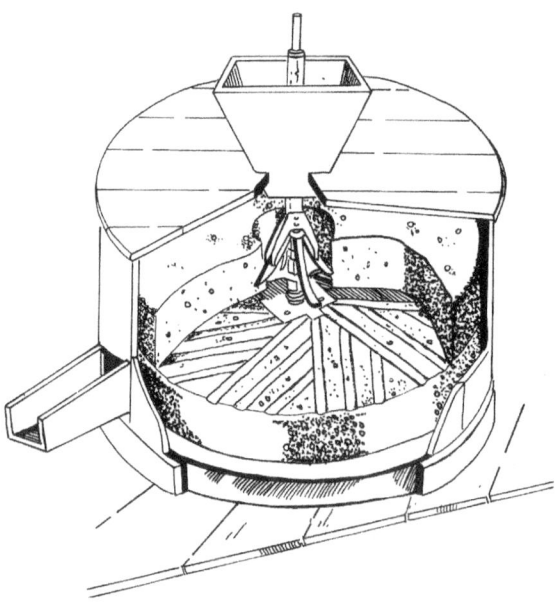

Figure 2. Cut-away drawing showing how millstones were mounted for grinding grain (illustration by Gary Adams, Winchester, Kentucky, reproduced with permission).

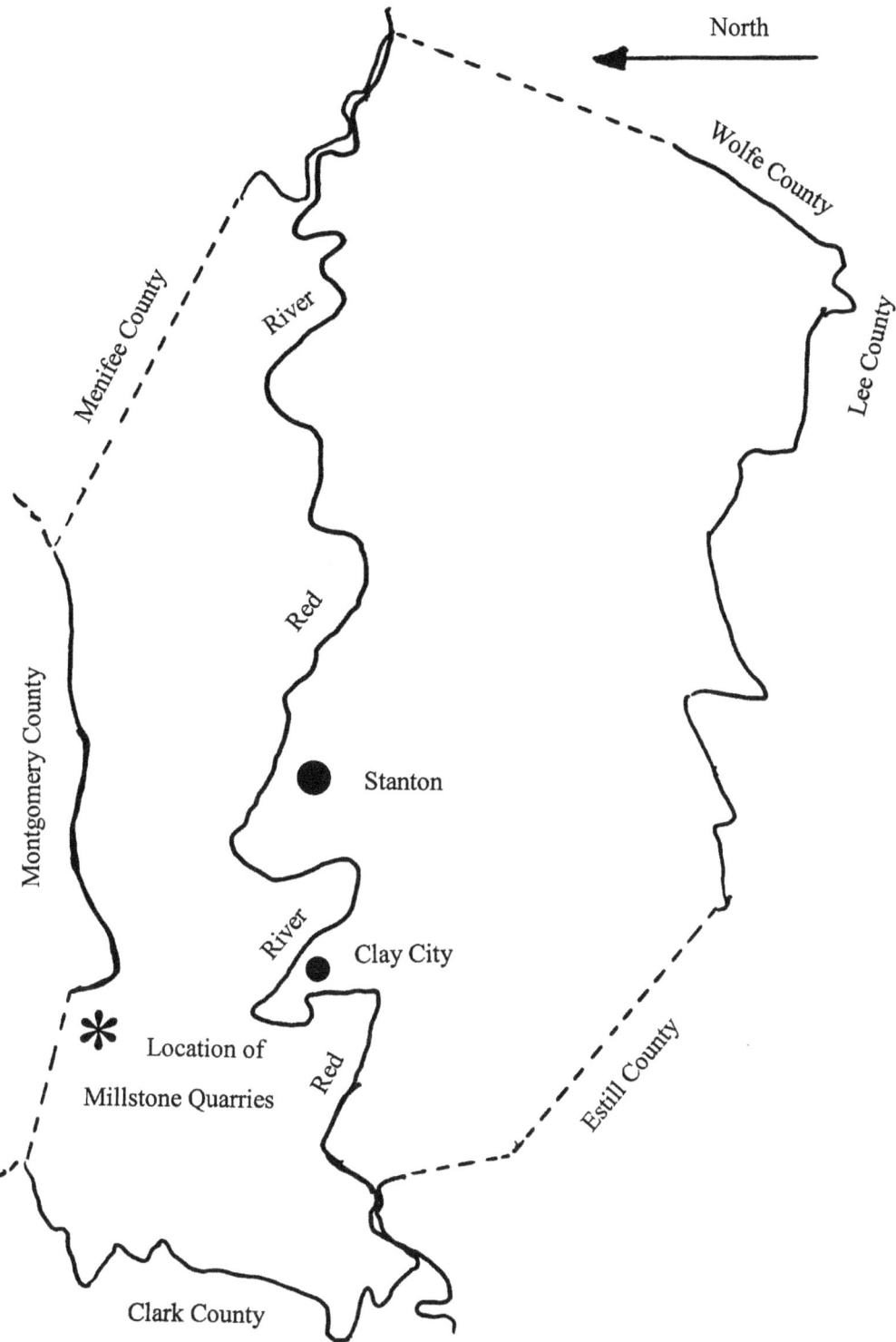

Figure 3. Map of Powell County, Kentucky, showing the location of the millstone quarries and other features. Drawing by Charles D. Hockensmith, Kentucky Heritage Council, Frankfort. Partially based on the "Geologic Map of Powell County, Kentucky" by Raymond Miller and Guy H. Briggs, Jr. (1929), Kentucky Geological Survey (used with permission of the Kentucky Geological Survey, Lexington, Kentucky).

the knob sides. Two of the quarries exploited in situ conglomerate deposits while the four remaining quarries utilized scattered boulders.

Before the introduction of steel roller mills in the late 19th century, high-quality millstones were essential for the operation of grist mills. Hughes (1869:91) noted that "as the *millstones* are the entire *key* which regulates the profits of the miller, we think much attention cannot be expended more profitably, than that bestowed in keeping them in proper order." Considering the key role that millstones played in the American milling industry, it is very fitting that archaeologists begin to document the quarries that produced these stones.

The Powell County millstone quarries are unique since they are well preserved and have been documented in detail. Currently, there are no other detailed archaeological studies of American millstone quarries. Very little fieldwork and research has been conducted on millstone quarries in the United States. Consequently, the Powell County quarries provide the first detailed view of American conglomerate millstone manufacture based on archaeological evidence. This book combines a study of the archaeological remains found at the quarries with a discussion of related archival records.

It is my desire that readers of this book will develop an appreciation for the conglomerate millstone industry in the United States. It is my hope that other archaeologists will be encouraged by this book to undertake similar studies of millstone quarries within the states where they reside. European archaeologists have made considerable strides in studying the millstone industries of several countries. A group of American scholars is likewise needed to focus their efforts on the millstone industry within the United States. As researchers work together, a more detailed synthesis of the American millstone industry will emerge.

1

An Archival Overview of the Kentucky Millstone Quarrying Industry

Very little has been written about the millstone quarries of Kentucky. Most available information consists of brief statements published in early geological reports and history books. Millstone quarries were located in Franklin, Letcher, Logan, Madison, McCreary, Marshall, Powell, Rockcastle, Whitley, and Woodford counties. Additional quarries may have existed in Logan County and many other counties that have adequate stone. Since the early geologists did not describe the millstone quarries, the industry may have been in decline by 1854 when the Kentucky Geological Survey was established. It is also possible that the early geologists felt that millstone quarrying was such a minor industry that they did not take time to report on it. Despite the limited information on quarry sites, there has been some interest in the millstones found in Kentucky (Crawford 1999; Dyche 1941, 1950; Webb 1933, 1935). The following discussion is organized by counties.

Franklin County

The firm of Miller, Railsback & Miller advertised their flint millstones from a quarry in Franklin County in the August 9, 1821, edition of *The Argus of Western America*:

LOOK HERE!

To Mill-Wrights and Mill-Builders.

THOSE who wish to purchase Mill-stones of the flint kind which have been cut for some years by Dudley and lately by Jeremiah Buckley, may now be had in Franklin county, Ky eight miles above Frankfort, on the river, where mill-stones are cut by the undersigned, on the land of Henry Miller, or near the adjoining land of M. Johnston. We have as good rock and as thick as necessary, and will cut as cheap & do our work as good as any. As the quality of these flint rock has been proven to be good for many years, it is not necessary to recommend them. We will deliver millstones any where in this state or in the states adjoining, by the purchaser paying the common cost of carriage and as our rock are not inferior to any of the kind, we are determined the workmanship shall be the same. We wish every one that wishes to purchase rock, to review them for themselves and in particular those who are real judges of quality and workmanship. All millstones proving not good cut by us either shall be cut gratis for the purchaser that are good, and of the same size. All letters to the undersigned will be attended to. They must be directed to Lawrenceburg, Franklin county, Kentucky. The prices of our Mill-stones are annexed, but the purchaser having cash can almost make his own bargins as we intend to work cheap, and are possessed with experience and a mechanical eye.

5 feet	$150.
4 feet	"100.
3 feet	"50.

And all sizes accordingly.

MILLER, RAILSBACK & MILLER

May 23rd, 1821.

Jeremiah Buckley also advertised his flint millstones in the November 8, 1821, edition of *The Argus of Western America*:

Mill Stone Quarry,

AT BUCKLEY'S FERRY, Franklin County. THE subscriber takes this opportunity of informing his friends and the public in general that he has on hand a good assortment of

MILL STONES,

and intends at all times to meet the calls of gentlemen that wish to get mill stones from him. As flint or grit is thought by the best of judges, to be superior to any other of the kind that has ever been discovered in the United States, for either corn or wheat when put in order, the French Burr not excepted it is hoped and sincerely requested by the subscriber that gentlemen in the west country wanting that article will inform themselves whether superiority of quality belongs to his Quarry or not; and for their information, he will refer them to a number of gentlemen who have his mill stones now in use: Col Robert McAfee, McCoum and Kennedy, Vandike and Keller, Robert Neal and Joseph Adams. These gentlemen have eight pair now in use on Salt River John Buford, Esq of Versailles and Mr. David Rice, on Clear Creek, Woodford county, Mr. James Rucker, of Caldwell county, one pair in use at this time, Mr. Gabriel Stansefer, on Main Elkhorn, has one pair now in use, also General George Baltzell and John Baltzell of this county, each one pair, Matthew Flourney, of Fayette, has also a pair in use. Mr. Flourney, Col McAfee and John Bufford, Esq. are at this time in the Legislature of our state, from whom, gentlemen living at a distance, through the medium of their representatives and others, may easily inform themselves.

The prices of my mill stones are as follows:

For Five Feet	$180
Four feet six inches	150
Four feet	125
Three feet nine inches	100
Three feet six inches	85
Three feet three inches	75
Three feet	60
Two feet nine inches	50
Two feet six inches	40

JEREMIAH BUCKLEY

October 18, 1821

Letcher County

A millstone quarry was located near Gordon, Kentucky, in Letcher County. This quarry utilized large conglomerate boulders that had rolled off Pine Mountain (Halcomb 1992). The Letcher County quarry began operation in the early 1800s and supplied local grist mills in this rugged area of southeast Kentucky (Halcomb 1992). During May 1992, Mr. Halcomb discovered that the quarry was flooded by a lake created by a recently constructed dam (Halcomb 1992).

Logan County

In 1816, Benjamin D. Price (South Union, Logan County) wrote to John L. Baker (New Harmony, Indiana) concerning millstones. Wrote Price (Arndt 1975:234), "I promised to write you when we could be satisfied in regard to the mill stones that are obtained in this country. We have got two pair of them at home and our people esteem them to be the best they have seen. There will be about six miles of land carriages to cumberland river which will make it convenient for you as they can conveyed by water to the place you want them."

Two years later (September 4, 1818), Joseph Allen (South Union) wrote Frederick Rapp (New Harmony) about providing millstones. Allen (Arndt 1975:565–566) said, "We have imployed a man to cut the four feet mill stones for corn, and as soon as it is practicable will send them to you, also the stone for your Hemp Mill with the other stones.... The man we Have imployed will undertake to git out as many pare as you want, if you find they will answer your purpose."

On February 25, 1819, Samuel G. Whyte (South Union) wrote Frederick Rapp to advise him that the millstones were finished and ready to be shipped (Arndt 1975:656). Benjamin S. Youngs (South Union) wrote to Frederick Rapp on May 20, 1819, to apologize for delays (Arndt 1975:712): "We have felt considerable concern for you on account of the Mill Stones, tanners tables & Hemp stone which you have expected from here. They have been all finished & taken to the bank of the Big Barren River about three months ago — but our Brethren have not been able as yet to get them a safe conveyance to Shawnee town...."

Samuel G. Whyte sent Frederick Rapp a bill for the millstones on September 1, 1819, and referred to them as "Goose Creek Mill Stones" (Arndt 1975:767).

The Goose Creek millstones were probably quarried in the general vicinity of South Union, Kentucky. The proposed shipping routes mentioned in the correspondence provide some clues. The first route mentioned 6 miles of land transport to the Cumberland River, which would have been to the south in Tennessee. The second reference mentions the Big Barren River (Barren River on later maps) which would have been to the north in Warren County. Using these references, the quarry could have been in Logan, Simpson or Warren County. The name "Goose Creek" does not occur as a named stream in that area today. Arnow (1984:283) mentioned that a pair of "Goose Creek burrs" were used in the Croft Mills of General James Winchester on Bledsoe Creek when it was sold in 1802. Bledsoe Creek is a tributary of the Cumberland River in southwest Russell County, Kentucky. This suggests that these millstones were being produced prior to 1802.

Madison County

Another early reference to a millstone quarry is McMurtrie's (1819:28) statement: "Between the head waters of Silver and Station Camp creeks, it runs into a beautiful breecia or pudding stone composed of primitive pebbles chiefly of quartz, a black schorl, that seldom weigh more than an ounce, and a cement formed of lime and sand: it is extremely valuable for millstones, numbers of which are annually manufactured from it, bearing a character for durability, seldom surpassed by any of the imported ones."

The above description probably refers to Millstone Ridge, which is located east of Silver Creek in southern Madison County. The geological map indicates that Millstone Ridge contains conglomerate deposits, but the millstone quarry has not yet been located. Conglomerate millstones have been observed in yards on the ridge system, and the conglomerate appears to be a good quality.

Marshall County

The first reference to a millstone quarry in far western Kentucky was in Loughridge's geological report on the Jackson Purchase region (Loughridge 1888:274): "The rock [conglomerate] is sometimes quarried and used for mill-stones, and seem to answer for the purpose very well, but [the stones] are gotten out and dressed with difficulty. A noted locality is known as Millstone Hill, on the north side of Jonathan creek, and about one and a half miles east of the Fair Dealing and Aurora road. The rock is a conglomerate of white and dark quartz or flint pebbles, and occurs in ledges two or three feet thick."

Loughridge's comments were repeated almost verbatim by Lemon (1894) in his book on Marshall County. An attempt was made to locate this quarry during February of 1992, but limited field survey of the vicinity and conversations with several older residents failed to yield any information about the quarry. The quarry may still exist, but the location is so vague that it could fall within several thousand acres of forested land.

McCreary County

A possible millstone quarry was reported to Charles Hockensmith by Tom Sussenbach (personal communication 1990) as being located along Indian Creek in McCreary County, Kentucky. Sussenbach was conducting an archaeological survey in the vicinity when the landowner mentioned the millstone quarry. He observed an unfinished millstone in Indian Creek (Sussenbach personal communication 1990). A plan to visit the quarry in January of 1991 did not occur. Fortunately, the author has a location for the quarry so that it can be investigated in the future.

Powell County

Several sources mentioned the millstone quarries of present day Powell County. Several previous studies by the author discuss the Powell County millstone quarries (Hockensmith 1988, 1990a, 1990b, 1993a, 1993b, 1994b, 1994a, 1994b, 1999b, 2002, 2003a, 2003b, 2004a, 2005a, 2006, 2008a; Hockensmith and Meadows 1996, 1997, 2006, 2007). Earlier in history, this area was located in both Clark and Montgomery counties. Since all the sources just listed refer to the Red River millstone quarries, they are included under Powell County.

The earliest reference to a Kentucky millstone quarry is in an interview with William Risk of Kidville in eastern Clark County. He stated, "I came up the next Fall, August 1793. The road was cut through at the same time, out to the stone quarry. Sq Road. Mill stones had been cut out in the Knobs, before I came out here" (John Shane, Draper MSS. 11CC 86). This reference indicates that millstones were being quarried in present day Powell County prior to 1793. Another early reference to the Powell County quarries was during Benjamin Logan's 1796 campaign for governor. Talbert (1962:284) stated that Logan "was traveling to the town of Winchester to address a gathering. He came upon one of his own wagons which he had sent to the Knobs region near Mount Sterling to obtain millstones. The wagon had been broken down or was mired down."

Two ads have come to light for the Red River millstones. First, the June 13, 1799, edition of the *Kentucky Gazette* contained an ad for the sale of five pairs of Red River millstones at Cleveland's Landing. Cleveland's Landing was located on the Kentucky River where the current I-75 bridge crosses in southern Fayette County (Larry Meadows, personal communication 2006). The second ad was published in the April 1, 1818, issue of the *Kentucky Reporter* and provided information on the sizes and prices of the millstones available. Both of these ads are quoted in their entirety in the chapter on "Kentucky Millstone Values."

Several lawsuits have been found that mention the Red River millstone quarries or millstone makers. An April 1, 1799, lawsuit between Valentine Huff and Peter DeWitt mentioned that DeWitt was a millstone cutter by trade living on Brush Creek (Clark County, Kentucky 1797). Fayette County mill owner John Higbee filed a lawsuit against Absolom Hanks of Clark County over a December 10, 1799, deal that required Hanks to deliver two millstones to Higbee from the Red River millstone quarry (Fayette County, Kentucky 1804). In 1803, James Daniel filed a lawsuit against Martin DeWitt concerning 25 grindstones to be cut by DeWitt in 1801 (Clark County, Kentucky 1803). Joseph Wilkerson sued Spencer Adams over a pair of millstones to be made at the Red River millstone quarry in 1807 (Clark County, Kentucky 1810). During 1822, the administrator of Martin Johnson's estate sued Spencer Adams over his failure to pay Johnson for making millstones at the Red River millstone quarry in 1819 (Clark County, Kentucky 1822). In 1826, Cornelius Summers filed a lawsuit against Spencer Adams concerning a pair of millstones at the Red River millstone quarry (Clark County, Kentucky 1826).

In 1861, Owen (1861:468) mentioned the Powell County millstone quarries in passing: "These knobs border the southern line of the county, and occasionally, when capped with the conglomerate, attain a considerable height, as is the case with the 'Pilot Knob,' between Black and Lulbegrud Creeks, remarkable for its millstone quarries."

In his 1884 geology report, Linney (1884:65) stated, "The great sandstone, often called the millstone grit, is the top rock on the highest part of Morris Mountain, Pilot, and Kash's Knobs. On the Clark Pilot, which is in Powell county, but near the Montgomery county line, it is one hundred and thirty-five feet thick, marked in nearly all its parts with rounded pebbles of white quartz.... This rock has been quarried on Pilot Knob and made into millstones, and they are highly prized under the name of hailstone grit."

Rockcastle County

Charles Colyer operated a millstone quarry in Rockcastle County, Kentucky, during the early 19th century. *The Argus of Western America*, Frankfort, carried an ad in the January 8, 1812, edition for a millstone quarry in Rockcastle County:

MILL STONES

I HAVE rented my QUARRY on Roundstone, about three miles from Rockcastle court house, to Mr. Colyear, who is now at work in manufacturing

MILL STONES,

Which I believe to be equal to any in America; at least they are equal to any that I have ever seen. Mr. Colyear proposes to sell on good terms, and take a small part in Whiskey.
Samuel Taylor.
January 7th, 1812.

The following year, Mr. Colyer published his own ad for the above millstone quarry in the October 9, 1813, edition of *The Argus of Western America*:

MILL STONES

THE subscriber has on hand a large and general assortment of MILL STONES, and intends keeping on hand all sizes of MILL STONES — which he will sell at his usual price — the Stone Quarry is within three miles and a half of Mount-Vernon, Rockcastle county, Kentucky.

N.B. — The MILL TONES [sic] are of a superior quality to any in this country, and are insured — and if they are not good, the money will be returned or another pair that shall be good.

CHARLES COLYER, JR.

The Filson Historical Society of Louisville, Kentucky, has an original handwritten contract between Charles Colyer and Sidney Payne Clay of Bourbon County (Colyer 1824). This document is from the Sidney Payne Clay Papers (A\C621a, folder 5) in The Filson Historical Society's Special Collections. The text of the contract and directions to Clay's house are quoted below with the permission of The Filson Historical Society:

28 Feb 1824

I Charles Colyer of Rockcastle County the state of Kentucky do hereby agree to deliver to Sidney P. Clay at his residence in Bourbon County on or before the 15 day of April next a pair of mill stone three feet four inches in diameter the runner to be eighteen inches thick through the eye and the bed stone about ten inches thick and do warrant the same to be good & sufficient. For which the said Clay does hereby bind himself his heirs &c. to pay said Colyer ninety dollars commonwealth money on the delivery of the stones in wittess [sic] whereof we have hereto let our hands and seals this 28th day of February 1824.

Sidney P. Clay (seal)
Charles Colyer

attst
George S. Shirley

A separate page attached to the contract provided directions to Sidney Payne Clay's house in Bourbon County and contained a receipt that the millstones were delivered:

Directions to where Sidney P. Clay lives. Keep the road from Lexington to Paris until you pass Bryan's station about two miles until you come to the Iron works road. Take the Right hand end of the Iron works road & in about seven miles you will come to Clay house near the side of the road.

April 20th 1824 Received of Sidney P. Clay Ninety dollars for a pair of mill stones of the size mentioned in within article.
Test
Will. Hearice [or Hearne]   John Wilson
Sidney p. Clay.s oblig-
-gation $ 90

Owen (1861:482) mentioned millstones being quarried in Rockcastle County during his geological survey conducted for 1858 and 1859: "The conglomerate member, which in the southeast, is 80 feet thick, thins out towards the head of Roundstone creek.... On Roundstone creek, six miles above its mouth, a quarry has been opened into this rock, which was formerly extensively worked for millstones."

Collins and Collins' (1874:376) *History of Kentucky* noted that parts of the eastern coalfield contained layers of millstone grit (conglomerate) which was used for millstones. In their discussion on Rockcastle County, Collins and Collins (1874:691) stated that "on Roundstone creek, 6 miles above its mouth, a quarry was formerly extensively worked for millstones." In his description of the geology of Rockcastle County, Sullivan (1891:18) provided the following description of the conglomerate deposits: "The Conglomerate formation, in the northern and central portions of this region, tops the limestone hills, and varies from 45 to 100 feet in thickness. In the southern part, the entire mountain above drainage consist of this formation, it being 150 to 250 feet thick. Much of the Conglomerate, in this region, consist of the "Hailstone Grit," the ledges of sandstone being very thick and the quartz pebbles large."

Whitley County

A millstone quarry was reported to be in the vicinity of Cumberland Falls in Whitley County, Kentucky. Reports of this quarry have been brought to Charles Hockensmith's attention by Tom Sussenbach (personal communication 1991) and Clarence Halcomb (1992). The cap rock at Cumberland Falls is a well cemented conglomerate and it is suspected that this material was utilized for millstones. The exact location of the quarry is presently unknown.

Woodford County

The Frankfort newspaper, *The Palladium*, carried an ad in the February 27, 1800, issue for millstones quarried in Woodford County, Kentucky:

MILL STONES.

A Fair and impartial trial of a pair of millstones cut in my quarry, of flint quality, of four feet in diameter, now running in Colonel Robert Johnson's mill on North Elkhorn Creek, was made with his Burr stones, as stated in his certificate, 330 lb. Of wheat being weighed into each mill, and the wheat weighing 59½ lb, per bushel, and both mills managed to the best advantage, and the flour bolted his superfine cloth immediately, that ground on the flint stones was first bolted, and made 243 lb. of flour, the Burrs 213 lb. leaving 30 pounds in favor of the flint stones, & of equal quality. I have here the Colonel's certificate for further information:

"I HAVE in my mill on North Elkhorn, a pair of French Burr Stones, four feet diameter; also, a pair the same size, which I purchased of Mr. John Tanner, cut in his quarry, in Woodford county. I have made trial of the flint mill stones I purchased of Mr. Tanner, and find them exceedingly good for manufacturing flour; and I further certify, that I am of opinion they are equal to any French Burrs of the same size in this state. Given under my hand, this 15th day of December, 1799."
   A true Copy.
  (Signed)   Robt. Johnson.

I Have for Sale, the Quarry out of which the above stones were cut, and five acres of ground including the same; and will make the terms cash (?) to the purchaser, by taking the price in mill stones and the produce of the country. Seven hands unto of whom may be warrent will make the 100 £ per month, clear of all expenses, at my usual selling prices, and the probability is, that the stones will bear a higher price than usual, if their value could be generally known, and if a speedy trial could be made of those who incline to purchase of me. I will warrant them to be equal to the Burrs. I have on hand, a few pair, from four feet two inches in diameter to four feet eight inches, which I will warrant as above.
       John Tanner

February 5th, 1800.

General Comments

In his discussion of limestone, geologist William Mather (1839:282) said of Kentucky millstones, "A stratum of silicious matter which varies in its texture from hornstone to a porous material like buhrstone, is found in many places in the cavernous limestone. The compact variety of this material was long used by the early settlers for flints for their rifles. Another form of it is used for hones. The coarser varieties are frequently seen of such a quality and texture, as to fit them for millstones. The localities of these materials are frequently called flint knobs, and their surfaces are thickly strewed with silicious masses."

During 1847, Lewis Collins (1847:158) published his *Historical Sketches of Kentucky*. In his description of the "Conglomerate Coal Series" in Kentucky, the following reference to millstones appears: "...a Conglomerate or pudding stone. It is composed of coarse pebbles of quartz, and fine grains of sand, rounded and cemented together by a silicious cement.... The rock is very firm, and is sometimes used for millstones to grind Indian corn."

Arnow (1983:392) in her *Seedtime on the Cumberland* made the following general comments on grindstones and millstones: "Grindstones were often cut from the harder sandstones, while a particularly hard, but not too coarse, conglomerate found above or between layers of sandstone could be shaped into millstones."

The Population Census records for 1860, 1870, and 1880 for Kenton County, Kentucky, mention individuals listed as millstone makers and millstone dressers. Nicholas Spanager, a 48-year-old Prussian born man, was listed as a millstone maker in the 1860 Census (Wieck 1983). The 1870 Kenton County Population Census listed both 25-year-old Herman Schulte and 58-year-old Nicholas Spanier (probably a different spelling of Spanager) as millstone dressers born in Prussia (Wieck 1986). The Kenton County Population Census for 1880 listed 34-year-old Herman Schulte (born in Prussia) as a millstone dresser (Wieck 1996). It is suspected that a number of other men involved in millstone making and dressing are listed in the U.S. Population Census records for other Kentucky counties. However, no one has undertaken the monumental task of searching all these records. The originals are handwritten and only available on microfilm. Fortunately, some individuals have transcribed and published this information for select counties and years.

2

The Environmental Setting of the Quarries

This chapter presents a very brief overview of the present-day environment of Powell County, Kentucky. Included are sections on the geology, soils, and flora. Discussions on climate and fauna were felt to be beyond the scope of the study.

Powell County is situated in east central Kentucky within the Knobs region. McFarlan (1943:194) described this region as "a narrow belt of country surrounding the Blue Grass, characterized by the presence of conical knobs, which are erosion remnants of the upland." The Red River, a tributary of the Kentucky River, flows westward across the center of the county.

Geology

The millstones from the Powell County quarries were manufactured from a conglomerate. Helton (1964:30) defines conglomerate as a "sedimentary rock which is usually composed of rounded quartz pebbles, cobbles, and boulders in a matrix of sand and finer material, and cemented with silica, iron oxide, or calcium carbonate."

The geological deposits of northern Powell County were essential for the establishment of the millstone quarries. Without these resources, local millstone production would have been impossible. The following paragraphs describe the conglomerate formations in the study area. In addition, these deposits are discussed in the context of Kentucky.

The conglomerate deposits on Rotten Point and Pilot Knob occur within the Lee Formation which dates to the Pennsylvanian geological period. The Lee Formation was described by McDowell (1978) as follows:

> Sandstone, quartzose, mostly conglomeratic, light-gray, weathers light brown; locally limonite stained; fine to medium grained; pebbles rounded white, gray, or pink quartz, mostly range from ⅛ inch to 1 inch across, most abundant in lower part of unit; pebbly beds about 125 feet thick on Pilot Knob; beds mostly a few inches thick or less, conspicuously crossbedded in sets several feet thick. Contains sparse casts of large plant remains as much as several feet long. Large angular chert fragments occur locally in lowermost few feet of unit on Rotten Point and ridges to the north; chert brownish-red to gray, fragments as much as 10 inches long. Unit forms cliffs and very steep slopes; basal contact sharp, commonly exposed where unit overlies limestone, covered where it overlies siltstone or shale; unconformity at base cuts as low as the upper part of the Cowbell Member of the Borden Formation on Pilot Knob and Rotten Point.

The conglomerate deposits in the study area represent a segment of a Pennsylvanian age channel deposit. Rice and Weir (1984:30) stated that these deposits "are thought to be remnants of a southwest-trending drainage system that extended from Ohio across central Kentucky into the Eastern Interior basin in western Kentucky." The authors described the formation of these channel fills as follows: "In the northern part of the belt, outcrops of older paleozoic sedimentary and metamorphic rocks yielded quartz sand and gravel that were carried to the southwest by rivers that drained the region. As the volume of sediment increased, the streams filled their channels, became braided, and spread out across their flood plains" (Rice and Weir 1984:43).

By the "Middle Pennsylvanian time, the remnants of the channels and their fills were buried by finer grained sediments from what was probably more easterly or southeasterly sources" (Rice and Weir 1984:42).

Today, the Pennsylvanian age channel deposits survive only as caps on the tops of knobs. Elsewhere, erosion has destroyed these ancient channel deposits. In Powell County, intact conglomerate deposits are restricted to the tops of a few knobs. Nearby, ridges and slopes sometimes contain large conglomerate boulders that may represent erosional remnants of these deposits. It is assumed that the more resistant conglomerate deposits dropped vertically as the softer underlying shales eroded away over thousands of years.

Soils

The soils of the knobs region (Colyer, Rockcastle, and Otway series) are shallow and not well suited to farming (Bailey and Winsor 1964:28). Bailey and Winsor (1964:28) stated that "the principal series within the valleys are the Tilsit on uplands, Monogahela on the terraces, Leadvale on the toe slopes, and Philo, Stendal, and Atkins on the bottomlands." A soil survey has yet to be published for Powell County. However, limited information on the soils near the quarries was contained in Fedders' (1983:23–27) report on the vegetation of the Spencer-Morton Preserve, which includes Pilot Knob. On the slopes of the preserve, Shelocta, Gilpin, and Rigley series soils are dominant (Fedders 1983:23). The top of the nearby Grape Knob contains Muse and Berks series soils (Fedders 1983:26). The crest of Pilot Knob is covered with Latham series soils (Fedders 1983:26). Whitley series soils were discovered along a small stream in the northwestern corner of the preserve (Fedders 1983:27). Given the close proximity of Pilot Knob to all the quarry sites and the similarity of the terrain, it is very probable that these soil types are common on most of the knobs in the area.

Flora

Powell County is within the Knobs Border Area of the Mixed Mesophytic Forest region (Braun 1950:118). Braun (1950:118) stated that a large proportion of the Knobs Border was originally occupied by "beech, tuliptree, basswood, red oak, buckeye, sugar maple, walnut, shellbark hickory, white ash, cucumber tree, red elm, bitternut hickory, white oak, redbud, dogwood, and other accessory species." She also noted that "the proportion of oaks and/or of chestnut increases on the upper slopes and, finally, near the summit of the ridge or knob, mixed mesophytic forest gives way to chestnut oak (Q. montana), to chestnut, to oak-hickory, to oak-chestnut-hickory, or to pine" (Braun 1950:119).

Fedders (1983:64) identified 39 tree species during his study of the Spencer-Morton Preserve which comprises 12 forest communities. These communities are chestnut-oak, chestnut oak–Virginia pine, chestnut oak–white oak, white oak, northern red oak–shagbark hickory, mixed oak-hickory, post oak–pignut-hickory, pignut hickory, sassafras, Virginia pine, and tulip poplar (Fedders 1983:68). The individual species comprising these forest communities are American beech, southern red oak, willow, red maple, silver maple, service berry, Hercules' Club, pignut, sweet pignut, shagbark hickory, white hickory, American chestnut, redbud, flowering dogwood, persimmon, white ash, black walnut, red cedar, sweet gum, yellow poplar, red mulberry, black gum, ironwood, sourwood, yellow pine, Virginia pine, cottonwood, wild black cherry, white oak, scarlet oak, black jack oak, chestnut oak, red oak, post oak, black oak, black locust, sassafras, white basswood, and slippery elm (Fedders 1983:64–66). Species of shrubs, vines and brambles at the Spencer-Morton Preserve include spicebush, wild hydrangea, blueberry, papaw, southern blackhaw, hawthorn, fragrant sumac, elderberry, American hazelnut, mountain laurel, poison ivy, sawbrier, dew berries, and rose (Fedders 1983:67).

3

Documenting the Quarries: Field Methods, Field Limitations, and Archival Research

The documentation of the Powell County millstone quarries presented a number of challenges and obstacles. Since there were no existing millstone studies to consult, the author was faced with determining what constituted quarrying remains and what remains merited recording. The quarry settings also introduced certain limitations. The following paragraphs describe field methods and discuss the limitations resulting from the field conditions.

Field Methods

Specific field methods were developed to locate and document quarrying remains. The initial step in documenting a millstone quarry was to systematically walk all ridge tops, slopes, and streambeds containing conglomerate boulders to locate evidence of quarrying. Shaped stones and stones containing drill holes were marked with red surveyors' flags. The next step was to clean the stones so that they could be documented. This usually involved sweeping or raking the thick leaf cover away from each specimen. In some cases, a trowel or shovel had to be used to remove soil partially covering specimens. Once all known specimens were exposed, a sketch map was prepared for each quarry. In the larger quarries, separate maps were prepared for individual ridges or concentrations. The author paced distances to obtain the measurements used in creating maps showing the approximate relationships between millstones, drilled boulders, excavated areas, and natural features. Distances were noted on the sketch maps and were later used to produce scaled maps on graph paper. More accurate maps utilizing a transit and stadial rod were not feasible for two main reasons. First, the author often worked alone and the manpower was not available. Second, the dense forest and steep slopes created obstacles that would have made such mapping extremely difficult.

From the beginning of the project, an attempt was made to standardize observations. The author developed a three-page form for recording millstones (see Appendix A) and another three page form was prepared for recording stones with drill holes (see Appendix B). Both forms have a similar introductory section which requests detailed information concerning the specimen's location, ownership, raw material description (texture, color, inclusions, etc.), recorder, photographs (roll and negative numbers), and date of fieldwork. The millstone form also recorded information on diameter, thickness, size of central hole (eye), tool marks, and the degree of completeness. In addition, an attempt was made to determine why the millstone was rejected and left at the quarry. The form for recording stones with drill holes collected specific information on each specimen (boulder, quarried slab, bedrock, etc.), length, width, thickness, number of drill holes (also whether cross-sectioned, complete or the base of the hole), the type of fracture produced when the stone was split (straight, concave, convex, irregular, or other), and whether tool marks were present. In addition, measurements (all in the metric system) were recorded for diameter, length, and orientation (vertical or horizontal) of each drill hole as well as the distance between the drill holes. Both forms were supplemented with sketches of these specimens. All the millstones and a sample of the drilled boulders were photographed with 35 mm black and white and color slide film. The millstones were photographed at different angles and distances to document their characteristics and their contexts. The photographic documentation also included other quarry remains (i.e. pits, benches, shaping debris) and general views of the quarries. The completed forms, field notes, photographs, and slides are currently curated at the Kentucky Historical Society's Special Collections in Frankfort.

Field Limitations

Field conditions at the millstone quarries created a number of problems that hindered the project. Perhaps the greatest problem was that of vegetation. To ensure the best visibility, fieldwork was undertaken in the early spring each year before the trees leafed out and weeds became a problem. The ground surface was obscured by a dense forest setting, which included various sizes of trees, bushes, vines, briars, and a thick leaf cover. Most rocks were at least partially obscured and the thinner rocks (or nearly buried boulders) were often completely covered by leaves. Because of manpower limitations, it was impossible to rake off the 10–15 cm thick layer of leaves covering the quarries or to clean off every low-lying rock. Also, thick layers of moss obscured some stones. It is also possible that some millstones were turned over by millstone makers or later sightseers resulting in the irregular base being exposed and thus obscuring the stone's true function. Consequently, a number of partially buried millstones or drilled boulders were undoubtedly overlooked at each quarry.

The second problem was specimens buried by erosion (colluviums) or by slumps. During past timber harvesting episodes, many stones were partially buried by soil eroding down slope. Since it was not feasible to completely expose each specimen, certain characteristics or tool marks may not have been visible. Several millstones were found almost completely buried along streams. While all stones showing evidence of being worked were exposed by limited excavation, it is probable that many other specimens were totally obscured by soil. Likewise, old slumps were observed on steep slopes that had buried sizeable areas with several feet of soil. Therefore, it is likely that slumps buried some of the millstones and drilled boulders at the McGuire and Toler quarries.

A third problem was that only the exposed portions of a specimen could be documented. Because of great weight and partial burial by soil, it was not possible to turn over the stones. Further, it was felt that such movement of a stone would alter its original context. Because of this limitation, many important flaws, tool marks, and other features were undoubtedly not recorded.

The very steep slopes in some quarry areas hindered the fieldwork. It was difficult to document a few specimens because of the extreme steepness of some slopes. There were a few instances when the author had to hold onto trees to keep from sliding down the slope. The steep slopes also made mapping difficult. Many horizontal map measurements were actually diagonal measurements on slopes. Some days, the author worked alone which limited the amount of documentation that could be accomplished.

A fifth limitation was that of lighting. As a result of being at the quarries from early in the morning to after sunset (during a four year period), differences in visibility were noted. For example, during periods of bright sunlight it was difficult to see drill holes. However, on very overcast days or just before dark, drill holes that were not noticed in good light became very prominent because of shadows. Due to lighting differences, it is felt that many boulders containing drill holes were simply overlooked during bright sunny days.

A final field limitation was the extended time required to become familiar with the range of millstone quarry remains. Since conglomerate is composed of quartz pebbles in a sandstone matrix, it has a very coarse texture. Consequently, a person's eyes must become accustomed to seeing evidence of breaking and shaping the stone. Also, tool marks are difficult to see on rough stone surfaces. Millstones that are nearly complete are clearly visible to most people. However, those millstones in very early stages of manufacture are difficult to distinguish when the ground is littered with similar looking boulders. The author noted that during each successive field season he was able to see more and more quarry remains. Upon returning to quarries that had been documented two years earlier, he recognized some crude early phase millstone blanks that he had initially overlooked. Consequently, the ability of a recorder to recognize the entire range of quarry remains increases tremendously as he or she spends more time in the field and as his or her eyes become accustomed to seeing minor modifications to conglomerate. This ability to recognize quarry remains also improves as a researcher develops a better understanding of the various stages in the manufacturing sequence.

Reflecting back on the fieldwork many years later, the author has considered that some of the millstones or perhaps their context may have been altered. When the quarries were initially documented, the author assumed that they were left much as they were when the last millstones were removed. Obviously, the quarries reverted to forest and were largely ignored because of the steep terrain and their remote location. However, some sources of potential alteration have come to mind. Some millstones may have been rolled over the steep slopes by the millstone makers to get them out of their way. Thus, these millstones may be in a secondary context. Other millstones could have been rolled down slopes or turned over by sightseers or locals during the past hundred years. Also, we know that these knobs were repeatedly logged to harvest the abundant timber crops. During the logging activities, large trees may have fallen on millstones or logs may have been pulled over some millstones. These activities could have broken or damaged some of the millstones. Other millstones may have been slightly moved when large trees were uprooted during storms. Finally, the forests in this area were once commonly burned to remove the undergrowth. Such fires could have caused some of the millstones and boulders to crack and spall.

Archival Research

The archival research consisted of three major components. The first research component consisted of collecting information on millstone quarrying. Information was compiled on millstone quarries in Kentucky, in the United States, and in Europe to establish a context for this study. Literature in several disciplines was consulted including the fields of archaeology, economic geology, mines and minerals, industrial archaeology, and history. University and state libraries in central Kentucky were utilized for sources. The geology library at the University of Kentucky was a vital repository for useful publications. Many resources were obtained through interlibrary loan. Also, dozens of letters were written to state historic preservation offices in the United States, to geological surveys in states known to be millstone producers, and to many other individuals thought to be sources of information. A lot of obscure information on American millstone quarries was obtained through this letter writing campaign. Finally, most of the European articles were discovered in response to letters written to industrial archaeologists in Great Britain and through contacts made by attending international millstone conferences.

The Internet has been a tremendous resource in recent years. Websites such as Making of America Books at the University of Michigan, Making of America periodicals at Cornell University, and HeritageQuest have made a tremendous amount of early literature available to researchers without leaving their offices. These websites can be searched by key words and all pages with those words can be examined on the computer screen. Thus, many sources mentioning millstones were found on these websites that would have gone undetected otherwise.

The second component of the archival research consisted of collecting information specifically related to the Powell County millstone quarries. This included the records in the courthouses of Powell, Clark, and Montgomery counties as well as public records in libraries. The records included deeds, lawsuits, county histories, census records, tax records, and genealogical records. Also visually scanned were the few surviving issues of local newspapers on microfilm at the Kentucky Historical Society including the *Mt. Sterling Whig* (1829), the *Kentucky Sentinel* (Montgomery County, 1865–1866), the *Sentinel Democrat* (Montgomery County, 1890), and the *Winchester Democrat* (1887–1888).

The early deeds and lawsuits were both difficult to find and equally difficult to use. These documents were handwritten and sometimes very faded. The early lawsuits were often tied in bundles along with many other lawsuits for the period. Individual lawsuits were tightly folded into small parcels that were composed of many sheets of paper of varying sizes. The staff of the Department of Libraries and Archives made photocopies of these lawsuits for us. Later, an effort was made to transcribe these lawsuits and include them as appendices. The nature of the handwriting and faded ink made it impossible to read each word. However, even with the missing information, we still have a good understanding of these lawsuits.

A third component of this research focused on collecting information on millstones from Kentucky sources. Through this research several ads were discovered that provided information on Kentucky millstone quarries, millstones for sale, and mills for sale that listed the types of stones. Several Frankfort newspapers were consulted including *The Argus of Western America* (1812, 1813, 1821), *The Palladium* (1800), and *The Commonwealth* (1839). Lexington newspapers checked included the *Kentucky Gazette* (1787–1820, 1825, 1835, 1840, 1867, 1870, and 1880), the *Lexington Observer & Reporter* (1831–1832, 1845, 1849–1850, 1855, 1858, 1865, 1867, and 1869–1870), and *Lexington Daily Press* (1875, 1885, and 1890).

4

An Historical Overview of the Powell County Quarries

A very brief overview of the Powell County millstone quarries and surrounding property was assembled from several sources. These sources mentioned the quarries, the sale of the quarries, lawsuits connected with the quarries, and other details. Through archival research many of the names of individuals who owned the land and worked the quarries have come to light. This research has been complicated since the county lines have shifted several times resulting in records' being in different courthouses for different periods of time. Gaps also exist because some of the courthouses burned during the Civil War, leaving no records of the quarries for those years. In spite of these limitations, much insight was gained through those documents that did survive. The presentation below is organized chronologically. Supplemental information was added when available. In the quoted material, references to the millstone quarries have been rendered in **boldface** to make them more readily apparent.

The earliest reference to millstone manufacture in the Powell County area (Powell did not become a separate county until 1852) was recorded in William Risk's interview (John Shane, Draper MSS. 11CC 86). A transcribed version of the interview was sent to the Kentucky Historical Society be Joseph R. Johnson (1965:1) of Lexington:

> WILLIAM RISK. 12.12 on the stone-quarry road, two or two and one-half miles from Kidville. Lived on Stoner near Col. Sudduth at the time of the taking of Morgan's Station. I came up here the year that the troups [*sic*] came out. That was in the Spring, I think (Waynes Troops) and I came up the next Fall, August 1793. The road was cut through at the same time, out to **the stone quarry**. Sq Road. **Mill stones had been cut out in the Knobs**, before I came out here. The road was out through the cane, coming down the hill to Stoner, six miles about, this side of Winchester, or about seven miles. This is called Big Stoner. Two miles nearer Winchester is a Branch called Little Stoner....
>
> The Stone Quarry Road when we came here at first only came to Big Stoner then, that I spoke of. There went a sort of road for a time — down a fork of Lullebegrud, along Comb's through where Mrs. Goff's place is. The road was afterwards opened as it now is.

Another early reference to the Powell County quarries was during Benjamin Logan's 1796 campaign for governor. Talbert (1962:284) states that Logan "was traveling to the town of Winchester to address a gathering. He came upon one of his own wagons which he had sent to the Knobs region near Mount Sterling **to obtain millstones**. The wagon had been broken down or was mired down."

In 1810, Joseph Wilkerson filed a lawsuit in Clark County Circuit Court against Spencer Adams concerning a broken plea of Covenant between Adams and James French (see Appendix D). On July 6, 1807, Adams agreed to deliver to French or his heirs one pair of millstones from the Red River Quarry in Clark County. Wilkerson was the assignee of William Palmer, who was the assignee of Owen Dolly, who was the assignee of Joel Tanner, who was the assignee of Moses Treadway, who was the assignee of James French. A pair of millstones of the best quality was to be made up to three feet in diameter and 17 inches thick within one month. Adams did not make the millstones and was being sued for $200 in damages.

On April 18, 1818, John Wilson of Philadelphia, Pennsylvania (through his attorney John McKenley of Lexington), sold 500 acres of land in Montgomery County to Spencer Adams, James Daniel, Peter DeWitt, Sr., and Moses Treadway of Montgomery and Clark counties for $200 (Montgomery County Deed Book 8, pages 364–365). This deed provides the following description of the 500-acre tract:

> ... tract or parcel of Land situated and being in the County of Montgomery on the waters of Red River containing five hundred acres by Survey and Bounded as follows.... Beginning at the Point of a Ridge near Black creek at two chestnut oaks thence N 72° W two hundred and fifty poles to a poplar, hickory and oak thence S 18° west three hundred & twenty poles to three Hickory saplings, thence S 72° E two hundred and fifty poles to three hickory saplings thence North 186° E three hundred & twenty poles to the Beginning including **the mill stone quarry** sold by said Wilson to said Adams & Anderson Pigg on the fifth day of July 1814 and is understood that said Adams and Daniel hold three fifths of said Land between them jointly and that said DeWitt & Treadway hold each one fifth....

Spencer Adams of Clark County sold a tract of land to Cornelius Summers of Montgomery County for $500 on October 10, 1823 (Montgomery County Circuit Court Records 1823). The 150-acre tract was located between the headwaters of Brush Creek and Black Creek in Montgomery County. This parcel was part of the 500-acre tract purchased from John Wilson by Spencer Adams, James Daniel, Anderson Pigg, and Peter DeWitt. The deed mentioned that it included "**the old and new Stone quarrys** to have and to hold the Same Tract and parcel of one hundred an fifty Acres of Land together with all and Singular the promises and appertainances there unto belonging or in any wise appertaining in fee Simple and forever."

During 1831, Singleton Davis sold John Ross 200 acres of land in Montgomery County on Black Creek that mentioned the stone quarry line. The land was paid for with a pair of three-foot-diameter millstones. Land was worth about $1 per acre at that time.

On August 27, 1835, Cornelius Summers and his wife Elizabeth of Montgomery County sold the 150 acres between the headwaters of Brush Creek and Black Creek to Anderson Pigg of Clark County for $25 (Montgomery County Deed Book 17). This was part of the 500-acre tract purchased together with Spencer Adams, James Daniel, Anderson Pigg, and Peter DeWitt. The deed mentioned that the tract included "**the old and new Stone quarrys.**" It is unknown why Summers would pay $500 for the tract in 1823 and then sell it for $25 in 1835. Perhaps Summers was able to recover his initial investment through the sale of millstones from the tract or was settling a debt with Anderson Pigg.

On May 16, 1853, Anderson Pigg filed a lawsuit against Moses Treadway's heirs in Powell County Circuit Court. In the lawsuit, "Anderson Pigg plaintiff states that about the year 1819 or 1820 he purchased from Moses Treadway then of Montgomery County a certain Boundary of land supported to contain one hundred acres and paid the same. The said land is a part of an original tract known as **the Mill Stone quarry tract** of five hundred acres. This land lies mostly if not all in Powell County, a portion of it however may lie in Clarke or Montgomery Counties — the new county line not being familiar to the plaintiff, it is a part of the same 500 acres upon which Benjamin Hedger now resides on the water of Black Creek."

Anderson Pigg indicated that he had "mislaid or lost the Bond" for the property and was suing Moses Treadway's heirs to obtain title to the land. The heirs named in the lawsuit were

John Treadway, Thomas Treadway, Peter Treadway, Christina Johnson (late Christine Treadway, wife of Isaac Johnson), William Treadway, Nancy Bennette (late Nancy Treadway, wife of Thomas Bennette), and Abigail Berry (late Abigail Treadway, wife of Newton Berry). On August 22, 1853, the Treadway heirs responded to Pigg's claims: "State that they know nothing of the purchase from their ancestor Moses Treadway, they admit they are the heirs and that Moses Treadway is dead. They denied that the plaintiff Anderson Pigg ever did purchase one hundred acres of **the stone quarry tract** from said Moses."

A court document dated August 24, 1853, suggests that Anderson Pigg won the case: "It is ordered that the defendants make to the plaintiff a deed for the land in the petition named on or before the 2nd day of next February." A partial transcription was published in a volume on Court records (Rogers 1998:37).

Peter D. Treadway and Margaret Treadway of Owen County, Indiana, appointed I. N. Berry as their attorney in fact on April 16, 1866, to bargain, sell and convey their share in Moses Treadway's land known as the "**Quarry tract**" of land (Powell County Deed Book 2, Page 393).

Nathan Adams and his wife Sonella Jane Adams of Powell County sold 180 acres to Nicholas Hadden of Montgomery County for $180 on November 26, 1867 (Powell County Deed Book 2, Page 394). The deed described the property as being "on the waters of Black Creek and Bounded as follows. Beginning the South East of **the Stone Quarry tract** of land and being that portion of the land of John Ross which he willed to his three Grand sons Thomas, William, and John Ross Containing one hundred and eighty acres."

A lawsuit filed in Powell County Circuit Court, dating between 1867 and April 1868, John A. J. Ross versus Benjamin Hedger, mentioned land on Black Creek with references to Nathan Adams and Anderson Pigg.

I. N. Berry (Attorney in fact) sold property to John H. Goff for $85 on April 6, 1868 (Powell County Deed Book 2, Pages 392–293). I. N. Berry was acting on behalf of Peter D. Treadway and his wife Margaret Treadway, Thomas Treadway, Nancy Bennett and Thomas Bennett, and Abigail Berry, wife of I. N. Berry. The deed noted that this was "the same tract of land known as **the Red River Mill Stone Quarry tract** of land situated in Powell County, Ky on the waters of Brush Creek tributary to the Red River." The deed further noted that this was "the tract of land of which Moses Treadway died, seized and possessed."

A document was filed in Powell County Circuit Court in 1870 that mentioned the millstone quarry tract (Rogers 1996:79–80). Harrison West had died in September of 1868, leaving his wife Mary E. West and infant daughter Levina L. West a tract of land on Black Creek. The land had been purchased from Alfred Nelson who took back about 25 acres for the unpaid balance. The remaining property was to be sold on May 16, 1870, for the benefit of Mary E. West and her daughter. The description of the tract mentioned "the center of the ridge to Wm. W. Nelson's line to a line in gap in the **stone quarry knob** to two old mark and line Chestnut oaks to J. E. Stewart's locust corner of the ridge near **the old Stone quarry hole**, and on to the beginning" (Rogers 1996:80).

Peter D. Berry and his wife Sarah Berry sold 250 acres to Thomas B. Ware on January 26, 1881, for $190 (Powell County Deed Book 4, Pages 151–152). The deed referred to "the ridge or what is known as **the Red River Stone quarry Knob**; thence in a southern direction with a marked line on the dividing ridge between Black Creek and Brush Creek to a pine corner in a survey on these premises in part of a survey recently made by John H. Goff."

On October 3, 1883, Thomas B. Ware and his wife Martha C. Ware sold Dillard P. Ware 150 acres for $400 (Powell County Deed Book 5, Pages 119–121). The deed mentioned "thence with said Wingate Young's line to his corner, a double pine on the ridge near what is known as **the Red River Stone Quarry Knob**; thence in a southern direction with a marked line on the dividing line between Black Creek and Brush Creek to Achilles Ware's corner."

John Ware to Samuel F. Flynn (Powell County Deed Book 32, Page 454). On March 21, 1942, one-half interest in a 400-acre tract was sold by Cora Ware Flynn and her husband Samuel F. Flynn to John Powell and Everett Powell for $1 (Powell County Deed Book 39, Pages 364–365). The deed mentioned "thence with said Wingate Young's line to his corner, a double pine on the ridge near what is known as **the Red River stone quarry Knob**; thence in a southern direction with a marked line on the dividing line between Black Creek and Brush Creek to Achilles Ware's corner."

Vina Powell sold half interest in a 400-acre tract to Everett Powell on July 12, 1948 (Powell County Deed Book 39, Page 593). Everett Powell and his wife Pinkie Powell sold James Marvin Clemons and his wife Wanda M. Clemons 400 acres for $1 (Powell County Deed Book 53, Page 84). The deed mentioned "thence with said Wingate Young's line to his corner, a double pine on the ridge near what is known as **the Red River stone quarry Knob**; thence in a southern direction with a marked line on the dividing line between Black Creek and Brush Creek to Achilles Ware's corner."

The property containing the Baker and Ware millstone quarries: James Marvin Clemons and his wife Wanda M. Clemons sold 400 acres to James D. Baker and wife Darlene H. Baker (one-fourth interest), Donald Baker and wife Mary Jo Baker (one-fourth interest), David Luzader and wife Sandra H. Luzader (one-fourth interest), and Ronnie Baker (one-fourth interest) for the sum of $25,000 on June 30, 1982 (Powell County Deed Book 88, Pages 196–199). The following excerpt from the deed description mentions the names of key people and locations (Powell County Deed Book 88, Page 197):

> Beginning at a post oak on the hillside, corner in the line of T. B. Ware's purchase of land from John H. Goff, between a graveyard and the house that was the residence of the T. B. Ware; thence a straight line nearly east to a small pine, corner on the ridge which divided the north and south forks of Brush Creek; thence with the meanders of said ridge, passing over what is known as the high divide on said ridge; to a chestnut oak, corner on the side of Pilot Knob; thence with a marked line nearly straight in an eastern direction to a hickory corner at the east point of Pilot Knob and in the line of Wingate Young; thence with said Wingate Young's line to his corner, a double pine on the ridge near what is known as **the Red River Stone Quarry Knob**; thence in a southern direction with a marked line on the dividing line between Black Creek and Brush Creek to Achilles Ware's corner; ... [detailed measurements included] ... corner to Hampton Heirs and Dora Berry; thence in a westerly direction to the east bank of Brush Creek; thence with the creek northwardly across the South branch of Brush Creek to the beginning, containing approximately four hundred (400) acres of land, be the same more or less.

As noted earlier in this chapter, the deeds provide some insight into the quarries through time. Unfortunately, there are many gaps that prevent our forming a detailed chain of title summaries for each modern tract with quarry remains. Hopefully, future research can fill these gaps and enhance our knowledge about the ownership of these quarries.

5

Families Associated with the Powell County Quarries

Archival research has identified several families with connections with the millstone quarries of Powell County. Some of these families (Adams, Daniel, DeWitt, Hanks, Johnson, Pigg, Stewart, Summers, and Treadway) are mentioned in documents that suggest that they were millstone makers. Other families (Adams, Berry, Daniel, Golf, Hedger, Nelson, Risk, Ross, Ware, and West) may have been owners of the quarries or nearby property at various points in time. A few individuals may have been land entrepreneurs that bought and sold property in the vicinity for profit. Additional names may represent later land owners or people that lived nearby. It is the purpose of this chapter to briefly discuss these individuals and what we know about them. In most cases, the available information is very limited. This information is presented alphabetically.

Adams Family

John A. Adams

John A. Adams owned land on Black Creek and Clay Lick behind Pilot Knob. John Adams was listed in Fayette County in June 1791 (Heinemann 1976:6). In 1800, there was a John Adams in Bourbon County and a John Adams in Clark County, Kentucky (Clift 1966:2). There were two John Adamses in Clark County in the 1810 U.S. Census and one in Montgomery County in the 1810 Census (Heritage Quest Online; Jackson and Teeples 1978a:4). During 1820, there were three John Adamses in Clark County and one in Montgomery County (Heritage Quest Online). The 1840 Clark County Census listed John Adams (Jackson and Teeples 1978a:2). In 1850, one John Adams was listed in Montgomery County and three were listed in Clark County (Jackson and Schaefermeyer 1976:2). The 1852 Tax Records for Powell County indicated that John A. Adams owned 404 acres of land (valued at $500) along the Lulbegrud Creek and had one white male over 21 years. During 1860, a John Adams was listed in Montgomery County (Jackson 1988a:6). These men may have just been property owners in the area around the quarries.

Nathan Adams

It is possible that John Adams had a son named Nathan; a Nathan Adams owned land on Black Creek, including the Toler quarry. It is clear from the records, however, that there was more than one Nathan Adams in the area. In 1800, there was a Nathan Adams in Scott County, Kentucky (Clift 1966:2). Nathan Adams was listed in the U.S. Census in 1810 as residing in Clark County (Heritage Quest Online; Hubble 1992:121; Jackson and Teeples 1978a:5). Miller (1971:3) noted that a family Bible in the possession of Clayton Rutledge of Clark County, Kentucky, recorded that a Nathaniel Adams was born July 2, 1796.

The 1852 Tax Records for Powell County indicated that Nathan Adams owned 100 acres of land (valued at $1,000) along the Red River. He had two slaves over 16 years and a total of 4 slaves (valued at $1,900). He owned three horses and mares (value of $50), and two mules ($100), and his total value was $2,340.

Nathan Adams was listed in the 1860 Powell County Census as a 35-year-old farmer born in Kentucky and single (Patrick 1981:19; Wonn 1981:22). He had $1,500 of real estate and $1320 of personal property (Wonn 1981:23). In 1870, a Nathan Adams lived in Clark County who was a 28-year-old farmer married to Sarah M. and had an 11-month-old daughter Patsy (Heritage Quest Online).

Payton Adams

Payton Adams lived at the head of Clay Lick Branch behind Pilot Knob. His only connection to the quarries may have been living close by. The records mention both a Payton Adams and a Peyton Adams, who appear to be different people. Payton Adams was listed in the 1840 U.S. Population Census for Clark County, Kentucky (Jackson and Teeples 1978c:2; Norris 1983:276).

Miller (1971:3) noted that a family Bible in the possession of Clayton Rutledge of Clark County, Kentucky, recorded that Payton Adams was born September 27, 1798. Adams (1976:26) mentioned a Peyton Adams of Clark County who was born September 11, 1801. In 1830, Peyton Adams was listed in Clark County, Kentucky (Jackson, Teeples, and Schaefermeyer 1976:1).

Peyton Adams was listed in the 1850 Clark County Census as a 52-year-old farmer born in Virginia and married to 45-year-old Polly Eubank Adams, born in Kentucky (Couey 1975:45). Susan Wallace and her children (Stephen P., 7, and James W., 6; born in Kentucky) lived with the Adamses and had $10,000 of property (Couey 1975:45). Payton Adams was also listed in the 1860 Clark County Census as a 61-year-old farmer born in Virginia and married to 54-year-old Polly, born in Kentucky (Norris 1981:99). No children were living in his home. He had $8,280 and $13,025 of property (Norris 1981:99). The 1870 Clark County Census listed a 71-year-old Payton Adams who was born in Virginia (Heritage Quest Online). Clark County cemetery records indicate that Payton Adams was born September 27, 1798, and died August 22, 1877 (Owen 1975:57). Peyton Adams died in 1838 (Adams 1976:26). Obviously, Payton and Peyton Adams are different people living in the same area. They were probably just local property owners.

Robert Adams

The 1820 Population Census for Clark Country listed a Robert Adams (Heritage Quest Online). His name was located two lines from Anderson Pigg's name, suggesting that they were neighbors. Robert Adams was listed in the 1850 Montgomery County Census as a 60-year-old farmer born in Virginia and married to 50-year-old Rose, born in Kentucky (Lawson 1986:48). Adams had three children at home (Anderson, 21; Luetha, 16; and Sarah, 13;

all born in Kentucky) and $300 of property (Lawson 1986:6). Robert Adams appears to be a land owner living near Anderson Pigg.

Spencer Adams

Spencer Adams was one of the key players in the millstone industry in Powell County, Kentucky. Spencer was born on February 23, 1773, in either Pittsylvania County or Culpeper County, Virginia, and died in Clark County, Kentucky, on June 1, 1863 (Adams 1976:25; Krauss 1983:57). He was the son of John Adams, Jr., and Sarah (Pigg) Adams and he married Sarah Corbin in Pittsylvania County, Virginia, on February 10, 1796 (Krauss 1982:57). A Spence Adams was listed in the Clark County, Kentucky, tax list for 1799 (T. L. C. Genealogy 1990:64). Adams (1976:25) noted that Spencer and Sarah Adams settled in Clark County about 1799. In 1800, Spencer Adams was listed in Clark County, Kentucky (Clift 1966:2). The U.S. Population Census for 1810 listed Spencer Adams in Clark County, Kentucky (Heritage Quest Online; Hubble 1992:121). An 1818 ad for the Red River Millstone quarry mentioned Spencer Adams and the sizes of millstones for sale (Kentucky Reporter 1818). The U.S. Population Census for 1820 listed Spencer Adams in Montgomery County, Kentucky (Felldin and Inman 1981:3; Heritage Quest Online; Lawson 1985:15). In 1830, Spencer Adams was listed in Clark County, Kentucky (Jackson, Teeples, and Schaefermeyer 1976:2). The records indicate that Spencer Adams was a millstone maker and one of the early quarry owners. The 1840 Estill County Census listed Spencer Adams (Jackson and Teeples 1978c:2).

Krauss (1982:57) noted that Adams was a millwright and farmer. Adams spent his last days living with Achilles S. Eubank and was buried on a farm near Winchester (Krauss 1982:60). Adams (1989) noted that "Spencer Adams, the son of John Adams, Jr., of Pittsylvania Co., Virginia, moved to Clark Co. in the late 1790s. Page 6 of the family papers states, 'Spencer Adams was a millwright by trade and he made mill rocks for the corn mills all over the state of Kentucky. He also operated a rock quarry where he mined and worked stones for building rocks.'"

It is also of interest that his son Elkanah Adams married Margery (also spelled Margerie) Treadway who was from another family involved in the millstone industry (Krauss 1982:57). Margerie Treadway (born 1802) was the daughter of Joel Treadway who was a younger brother to Moses and John Treadway (Krauss 1982:63). Spencer Adams and Anderson Pigg were mentioned together on a court document concerning the estate of Elkanah Adams on January 5, 1807 (Krauss 1982:61).

Krauss (1982:58–61) has listed several early documents related to Spencer Adams. He shows up in deeds and other local records. Adams bought and sold several tracts of land between 1799 and 1831 in Clark and Montgomery counties. Some of these documents are pertinent to this study. Krauss mentioned three sources related to the millstone quarry (Krauss 1982:59):

> Montgomery Co., Ky. Records —1814 ... Spencer Adams of Clark Co., Ky. on 5 July 1814, purchased a tract of land containing a millstone quarry in Montgomery County, Ky. This was purchased together with Anderson Pigg (no doubt a relative) from John Wilson of Philadelphia, Pa. by John McKinley, his attorney in fact of Fayette Co., Ky. This mill or quarry was situated on the waters of Red River....

> Montgomery Co., Ky. Deed Records —1816 ... Spencer Adams of Clark Co., Ky. on 18 Apr. 1818, together with James Daniel, Peter DeWitt Sr., & Moses Treadway purchased from John Wilson of Philadelphia, through John McKinley his attorney of Fayette Co., Ky. a tract of land in Montgomery Co. situated on the waters of Red River and containing 500

acres. (Spencer's son Elkanah Adams married a niece of Moses Treadway, Margerie Treadway, dau. of Joel Treadway)....

Montgomery Co., Ky. Records—1823 ... Spencer Adams of Clark Co., on 10 Oct. 1823, sold a tract of land to Cornelius Somers of Montgomery Co., for $500. This was part of the tract purchased together with James Daniel, Anderson Pigg, and Peter DeWitt from John Wilson. It was situated between the head waters of Brush Creek and Black Creek. Witness, L-yleton G. Davis.

Berry Family

Benjamin Berry

Benjamin Berry was mentioned as a co-owner of land on Lulbegrud Creek as early as 1782 (Weaks 1925:24). On June 8, 1782, Articles of Agreement for the division of 5,800 acres of land on Lulbegrud Creek mentioned Benjamin Berry, Marquis Calmes, Sr., Marquis Calmes, Jr., Benjamin Combs, and Cuthbert Combs (Weaks 1925:24). Berry was included in the 1788 tax list for Fayette County as owning five horses (Todd et al 1965:1). Benjamin Berry was also listed in the Fayette County tax list for 1789 (Heinemann and Brumbaugh 1938:11). Rogers (1986:31) indicated that Benjamin Berry was born October 16, 1724, and married a lady named Elizabeth.

Rogers (1986:30) provided evidence that Berry owned land near Lulbegrud Creek which flows through present day Clark, Montgomery, and Powell counties near the millstone quarries: "The earliest available record of a Pioneer of the name of Berry, was that of a certificate for 1000 acres of land granted by the Virginia Commission to Benjamin Berry giving location as between Lulbegrud and Hinskon in 1775. Benjamin Berry gave Power of Attorney in 1798 to John McGuire."

Benjamin Berry was listed in Fayette County on August 4, 1789 (Heinemann 1976:11). In 1800, Benjamin Berry was listed in Bourbon County, Kentucky (Clift 1966:21). Benjamin Berry was included in the 1820 Census for Fayette County (Felldin and Inman 1981:22). In 1830, Benjamin Berry was listed in Fayette County, Kentucky (Jackson, Teeples, and Schaefermeyer 1976:14). Berry may have been one of the early millstone makers but at very least was an early land owner in the area.

Isaac Newton Berry

Isaac Newton Berry was born in 1820 (Rogers 1986:44). He was the son of Thomas Berry and Sally DeWitt Berry and the brother of Peter DeWitt Berry (Rogers 1986:44). Berry married Abigail Treadway (Rogers 1986:44).

Isaac N. Berry was listed in the 1850 Clark County Census as a 30-year-old farmer ($600 of property) born in Kentucky and married to 36-year-old Abigail, born in Kentucky (Couey 1975:35). Berry had 29-year-old Peter Berry living with him who may be a younger brother (Couey 1975:35). The 1852 Tax Records for Powell County indicated that Isaac N. Berry owned 134 acres of land (valued at $170) along the Brush Creek, had one white male over 21 years, owned two horses and mares (value of $60) and 7 cows ($44), and had a total value of $774.

Isaac Berry was listed in the 1870 Powell County Census as a 49-year-old man and married to 56-year-old Albie (Patrick 1988:5). He had $300 of property (Patrick 1988:5). Berry owned land in the vicinity of the millstone quarries and was related to the millstone making DeWitt family through his mother.

Peter DeWitt Berry

Peter DeWitt Berry was born in 1821 and died in 1904 (Rogers 1986:44). He was the son of Thomas Berry and Sally DeWitt and the brother of Isaac Newton Berry (Rogers 1986:44). Berry married Sadie Ewen who was born in 1826 and died on January 17, 1903 (Rogers 1986:44).

The 1852 Tax Records for Powell County indicated that Peter D. Berry owned 100 acres of land (valued at $800) along the Brush Creek, had one white male over 21 years, and owned one horse (value of $40) and eight cows ($44).

P. D. Berry was listed in the 1860 Powell County Census as a 38-year-old farmer born in Kentucky and married to 28-year-old Sarah, also born in Kentucky (Patrick 1981:17; Wonn 1981:20). Berry had four children (John T., 6; Newton, 4; Arminta, 3; and William, 1; all born in Kentucky), and $500 and $300 of property (Wonn 1981:20). Peter Berry was listed in the 1870 Powell County Census as a 48-year-old man and married to 45-year-old Sarah (Patrick 1988:5). Berry had seven children living in his home (John T., 16; Isaac, 14; Arminta, 12; William, 11; George R., 9; Franklin, 7; and James, 4). He had $2,000 and $600 worth of property (Patrick 1988:5). In 1880, Peter D. Berry was listed in the U.S. Population Census for Powell County as a 59-year-old farmer living with his 53-year-old wife Sarah and children Arminta (22); William (20), George R. (19); Franklin (18); James (14), and John T. (25) (Morton 1994:5; Patrick 1994:5). Peter DeWitt Berry, like his brother Isaac Newton Berry, was related to the millstone making DeWitt family through his mother.

Thomas Berry

Thomas Berry was listed in Fayette County on February 27, 1790 (Heinemann 1976:11). Thomas Berry, Sr., and Thomas Berry, Jr., were both listed in the Clark County, Kentucky, tax list for 1793 (T. L. C. Genealogy 1990:2). According to Rogers (1986:52), Thomas Berry was born in 1790, married Sally DeWitt (born 1805), and had several children. Thomas Berry was listed in the Clark County, Kentucky, tax lists for 1794, 1795, 1796, 1797, and 1799 (T. L. C. Genealogy 1990:11, 23, 39, 55, 64). In 1800, two Thomas Berrys were listed in Clark County, Kentucky (Clift 1966:22). Three Thomas Berrys were listed in the 1810 Census for Clark County: Thomas Berry, Thomas H. Berry, and Thomas Berry, Jr. (Heritage Quest Online; Hubble 1992:122). In the 1820 Census for Clark County, three Thomas Berrys were also listed (Felldin and Inman 1981:22; Heritage Quest Online). In 1830, Thomas Berry was listed in Clark County, Kentucky (Jackson and Teeples 1978b:14). Two Thomas Berrys were listed in the 1840 U.S. Population Census for Clark County, Kentucky (Norris 1983:287, 291). In 1822, Spencer Adams sued Thomas Berry for attacking his character and good name (Clark County, Kentucky 1822).

Thomas Berry was listed in the 1850 Clark County Census as a 55-year-old farmer born in Virginia and married to 45-year-old Sarah DeWitt Berry, born in Kentucky (Couey 1975:35). Berry had four children at home (Louisa, 16; Joseph, 13; Amanda, 11; and Joicy, 8; all born in Kentucky) and $4,000 of property (Couey 1975:35).

The 1852 Tax Records for Powell County indicated that Thomas Berry, Sr., owned 500 acres of land (valued at $800) along the Brush Creek. He had one white male over 21 years; three horses and mares (value of $70); one mule ($50); 13 cows ($130); and a total value of $2,340. He also had five children between 6 and 18 years of age.

Thomas Berry was listed in the 1860 Powell County Census as a 70-year-old farmer born in Virginia and married to 53-year-old Sarah, born in Kentucky (Patrick 1981:17; Wonn 1981:20). Berry had three daughters and one son at home (Loueza, 23; Amanda, 22; Jincy, 20; and Joseph, 21; all born in Kentucky) and $1,200 and $300 of property (Wonn 1981:20). This appears be the Thomas Berry who was born in Virginia in 1790 and died in Powell County in 1861 (Rogers 1986:43). He was the son of Thomas Berry (born in 1760) and Margaret Newton Berry and married Sally DeWitt (born 1805), the daughter of Peter DeWitt, Jr. (Rogers 1986:44). His father Thomas Berry (1760–1834) married Patsy Morehead on February 15, 1815, and died in Clark County, Kentucky (Rogers 1986:43). Thomas Berry was related to the DeWitt family that made millstones.

Daniel Family

Beverly Daniel

Beverly Daniel was born about 1765 and died in Clark County, Kentucky, in 1827 (Heinemann 1934:239). He married Sarah Daniel first, and after her death, he married Esther Hampton in Clark County on August 14, 1813 (Heinemann 1934:239). Esther died September 21, 1861 (Heinemann 1934:240).

Beverly Daniel was listed in the Clark County, Kentucky, tax lists for 1794, 1795, 1796, 1797, and 1799 (T. L. C. Genealogy 1990:13, 25, 41, 56, 65). In 1800, Beverly Daniel was listed in Clark County, Kentucky (Clift 1966:72). The U.S. Population Census for 1810 listed Beverly Daniel in Clark County, Kentucky (Heritage Quest Online; Hubble 1992:127). The U.S. Population Census for 1820 listed Beverly Daniel in Montgomery County, Kentucky (Felldin and Inman 1981:71; Heritage Quest Online). In 1830, Beverly Daniel was listed in Montgomery County, Kentucky (Jackson and Teeples 1978b:46). The 1840 Montgomery County Census listed Beverly Daniel (Jackson and Teeples 1978c:56). Finally, Beverly Daniel was listed in the 1850 Montgomery County Census as an 89-year-old blind man born in Virginia (Lawson 1986:37). He was living with James M. Daniel.

James Daniel

In 1800, two James Daniels were listed in Clark County, Kentucky (Clift 1966:72). The U.S. Population Census for 1810 listed James M. Daniel, James Daniel, Sr., and James Daniel, Jr., in Clark County, Kentucky (Heritage Quest Online; Hubble 1992:127). An 1818 ad for the Red River Millstone quarry mentioned James Daniel as living in Winchester (Kentucky Reporter 1818). The U.S. Population Census for 1820 also listed James Daniel in Clark County and James M. Daniel in Montgomery County, Kentucky (Felldin and Inman 1981:71; Heritage Quest Online). U.S. Population Censuses for Montgomery County during 1830 (Lawson 1985:28) and 1840 (Lawson 1985:48) listed James Daniel. One James Daniel married Ophelia Parker before 1824 (Boyd 1961:25).

James Daniel was involved in a lawsuit against Martin DeWitt in Clark County Circuit Court in 1803 (Clark County, Kentucky 1803). Daniel sued DeWitt over lack of payment for 25 grindstones, which were "to be eighteen inches in diameter and four inches thick to be good merchantable gritt to be cut in a good workman like manner to be paid on or before the first day of March next." The agreement was made on November 16, 1801.

James M. Daniel was listed in the 1850 Montgomery County Census as a 62-year-old shoemaker born in Virginia and married to 25-year-old Mary, born in Kentucky (Lawson 1986:37). Daniel had two children at home (William H., 7; Sarah A., 3; both born in Kentucky) and $500 of property (Lawson 1986:37). James M. Daniel was listed in the 1870 Powell County Census as a 78-year-old man and married to 43-year-old Mary (Patrick 1988:9). Daniel had four children living in his home (Mary E., 19; William, 26; Betsy, 30; and Newton, 12). He had $500 worth of property (Patrick 1988:9). A James Daniel was listed as being buried in a Clark County cemetery; he was born April 19, 1852, and died October 2, 1902 (Owen 1975:27). Obviously, this is not the James Daniel mentioned in the early records, but he could be a relative. The earlier James Daniel was involved in the Red River millstone quarries.

DeWitt Family

Martin DeWitt

Martin DeWitt was listed in Fayette County on February 27, 1790 (Heinemann 1976:28). Martin DeWitt was also listed in Fayette County on July 22, 1789, and February 27, 1790 (Heinemann 1976:28). The Clark County, Kentucky, tax lists for 1793, 1796, and 1799 listed Martin DeWitt (T. L. C. Genealogy 1990:4, 42, 65). DeWitt was listed in the 1797 and 1799 tax records for Montgomery County, Kentucky (Dunn 1996a:22, 30). In 1800, Martin DeWitt was listed in Clark County, Kentucky (Clift 1966:79). The 1810 U.S. Population Census for Clark County listed Martin DeWitt (Heritage Quest Online; Hubble 1992:127). DeWitt made grindstones and may have made millstones.

Peter DeWitt, Sr.

Peter DeWitt, Sr., was born on July 8, 1753, and he married Jane Bray in July of 1774 (Bowman 1963:7; Rogers 1986:51). Peter DeWitt was listed in Fayette County on July 22, 1789, February 27, 1790, and November 5, 1792 (Heinemann 1976:28; Heinemann and Brumbaugh 1938: 11). The Clark County, Kentucky, tax lists for 1793, 1794, 1796, 1797, and 1799 listed Peter DeWitt (T. L. C. Genealogy 1990:4, 13, 42, 57, 65). He was also listed in the 1797 and 1799 tax records for Montgomery County, Kentucky (Dunn 1996a:22, 30). A February 10, 1797, lawsuit (Harrison verses DeWitt) listed on the outside "Peter DeWitt the millstone cutter by trade living on Brush Creek." In 1800, Peter DeWitt was listed in Clark County, Kentucky (Clift 1966:79). The 1800 tax records for Montgomery County, Kentucky, indicated that Peter DeWitt owned 100 acres on Lulbegrud and had two horses (Dunn 1996a:55). The U.S. Population Census for 1810 listed Peter DeWitt in Clark County, Kentucky (Heritage Quest Online; Hubble 1992:127). Peter DeWitt, Sr., died April 22, 1834 (Rogers 1986:51). His oldest son, Peter DeWitt, Jr., was married to Massey DeWitt (Rogers 1986:51). In 1822, Spencer Adams sued Peter DeWitt for attacking his character and good name (Clark County, Kentucky 1822).

It is interesting to note that the 1790 and 1820 U.S. Population Censuses listed a Peter DeWitt in the Marbletown area of Ulster County, New York (Heritage Quest Online), which was the major area of conglomerate millstone manufacture. Likewise, the 1800 U.S. Population Census listed a Peter DeWitt in Fayette County, Pennsylvania (Heritage Quest Online), another area of millstone manufacture. It is possible that the DeWitt family had some involvement with the millstone industry in New York and Pennsylvania. The U.S. Population Census for 1810 listed Peter DeWitt in Clark County, Kentucky (Heritage Quest Online). Rogers (1986:51) stated, "The New Jersey DeWitts had originally come from New Amsterdam (Holland) then on into New York."

The daughter of Peter DeWitt, Sr., Catherine DeWitt, married Moses Treadway (Bowman 1963:7; Krauss 1982:63). William Risk's interview (John Shane, Draper MSS. 11CC 86) noted that "Peter Dewit lived out in the Knobs, on Brush, when I came. He died in these parts and his wife moved to Ohio."

Thomas DeWitt

There was a Thomas DeWitt in the Kingston Township of Ulster County, New York, listed in the 1790 U.S. Population Census (Heritage Quest Online). Likewise, a Thomas DeWitt was listed for Kingston in Ulster County, New York, in the 1800 U.S. Population Census (Heritage Quest Online). During 1810 and 1820 Population Census schedules, there is a Thomas DeWitt listed in Warwasing in Ulster County (Heritage Quest Online). The Thomas DeWitt in Clark County, Kentucky, was mentioned much later in a lawsuit and may be a descendent of Peter.

William DeWitt

In the 1790 U.S. Population Census, both a William DeWitt and a William A. DeWitt were listed in the Rochester Township of Ulster County, New York (Heritage Quest Online). There were four William DeWitts in the 1800 U.S. Population Census for Ulster County, New York: William (Rochester), William (Shawagunk), William A. (Marbletown), and William, Jr. (Rochester) (Heritage Quest Online). During 1810, the Population Census schedules listed two William DeWitts in Ulster County: William (Shawagunk) and William (Hurley). By 1820, one William DeWitt was listed in the U.S. Population Census at Shawagunk in Ulster County (Heritage Quest Online). The William DeWitt in Clark County, Kentucky, was mentioned much later in a lawsuit and may be a descendent of Peter.

Golf Family

John H. Golf

John Hodges Golf married Anna Prewitt on April 9, 1857 (Boyd 1961:40). In 1850, John Golf was listed in the Clark County Census (Jackson and Schaefermeyer 1976:159). John H. Golf was listed in the 1860 Clark County Census as a 40-year-old farmer and married to 30-year-old Patsy (Norris 1981:126; Jackson 1988a:398). Golf had six children living in his home (Thomas, 11; Henrietta, 9; Levi, 7; Strawder, 5; Emily, 3; William, 2, all born in Kentucky), and 65-year-old Nancy who was probably his mother (Norris 1981:126). He had $88,020 and $32,627 worth of property (Norris 1981:126). In 1870, John Golf was living in Clark County (Jackson 1988b:237). John Hedges Golf (May 9, 1821 to May 23, 1901) was buried in Winchester in Clark County (Owens and Couey 1983:40). Obviously, there were at least two John H. Golfs who may have been property owners in the vicinity of the millstone quarries.

Hanks Family

Absalom Hanks

Absalom Hanks was listed in the Clark County, Kentucky, tax lists for 1793, 1794, 1795, 1797, and 1799 (T. L. C. Genealogy 1990:5, 15, 27, 58, 67). In 1800, Absalom Hanks was listed in Clark County, Kentucky (Clift 1966:123). He was also listed in the 1810 U.S. Population Census for Clark County, Kentucky (Heritage Quest Online). In 1830, Absalom Hanks was listed in Green County, Kentucky (Jackson, Teeples, and Schaefermeyer 1976:78).

Baber (2004:504–505) indicated that Absalom Hanks, Sr., was the son of Peter Hanks III of Maryland. The date of his birth is uncertain and thought to be either 1760 or 1773 (Baber 2004:504). Hanks married Malinda Tribble and had four children: Absalom, Jr., Mary, Elizabeth "Betsey," and Margaret (Baber 1959:268, 2004:505). The Hankses "resided at the N. side of Winchester, in the fork where the road branches off NE to Mt. Sterling" (Baber 2004:505). Absalom Hanks died in Clark County, Kentucky, in 1827 (Baber 1959:268, 2004:505).

Absolom [sic] Hanks was sued by John Higbee of Fayette County in 1804 (see Appendix C) for his failure to deliver a pair of millstones from the Red River quarry to Higbee's mill near Lexington (Fayette County, Kentucky 1804). Currently, it is not known whether Hanks was a millstone maker or was acting as a middleman in selling millstones.

Hedger Family

Benjamin Hedger

Benjamin Hedger married Matilda Lloyd on September 5, 1827 (Boyd 1961:48). In 1830, Benjamin Hedger was listed in Estill County, Kentucky (Jackson, Teeples, and Schaefermeyer 1976:83). The 1850 Census for Montgomery County listed Benjamin Hedger (Jackson and Schaefermeyer 1976:190). The 1852 Tax Records for Powell County indicated that Benjamin Hedger owned 150 acres of land (valued at $500) along the Black Creek. He had one white male over 21 years, one horse (value of $50), and a total value of $550. He also had four children between 6 and 18 years. During 1860, the only Benjamin Hedger was listed in Anderson County (Jackson 1988a:469).

Benjamin Hedger was listed in the 1870 Powell County Census as a 62-year-old man born in West Virginia and married to 38-year-old Elizabeth, born in West Virginia (Patrick 1988:3–4). Hedger had a daughter, Lavina (17 years old), born in Kentucky. Hedger had $2,000 and $300 of property (Patrick 1988:3). Apparently, Hedger was just a local land owner.

Johnson Family

Martin Johnson

A Martin Johnson was listed in the 1790 U.S. Population Census in Queens, New York, in the Jamaica Township (Heritage Quest Online). It is unsure whether this is the Martin who moved to Kentucky. Martin Johnson was listed in the Clark County, Kentucky, tax lists for 1793, 1794, 1795, 1796, 1797, and 1799 (T. L. C. Genealogy 1990:6, 16, 29, 46, 59, 67). In 1800, Martin Johnston was listed in Clark County, Kentucky (Clift 1966:155). He was listed in the U.S. Population Census for Clark County, Kentucky in 1810 (Hubble 1992:123). In 1830, a Martin Johnson was listed in Henry County, Kentucky (Jackson, Teeples, and Schaefermeyer 1976:98). Martin Johnson was listed in the 1840 U.S. Population Census for Clark County, Kentucky (Jackson and Teeples 1976:119; Norris 1983:267). This later Martin Johnson may be a son of the older Martin who died before 1821.

Martin Johnson was working as a millstone maker in Clark County in 1819. His son William Johnson filed a lawsuit against Spencer Adams on July 19, 1821 (see Appendix E). According to the lawsuit, Martin Johnson worked for Spencer Adams in 1819 and did not receive the agreed upon payment for his part in making millstones (Clark County, Kentucky, 1821). Since the lawsuit mentioned that Martin Johnson was an old man and a good hand in the millstone business, it is probable that he was involved in earlier quarrying activities, possibly in New York.

Nelson Family

William Nelson

In 1800, William Nelson was listed in Montgomery County, Kentucky (Clift 1966:215). William Nelson was listed in Montgomery County in 1810 (Jackson and Teeples 1978a:574). A William Nelson was listed in the 1820 U.S. Population Census as living in Montgomery County, Kentucky (Felldin and Inman 1981:207; Heritage Quest Online). In 1830, two William Nelsons were listed in Montgomery County, Kentucky (Jackson, Teeples, and Schaefermeyer 1976:134). The 1840 Montgomery County Census listed William Nelson (Jackson and Teeples 1978c:162). William Nelson (November 21, 1797, to December 25, 1857) was buried in a cemetery at Grassy Lick about five miles from Mt. Sterling in Montgomery County (Boyd n.d.).

William H. Nelson was listed in the 1850 Montgomery County Census as a 46-year-old farmer born in Virginia and married to 47-year-old Charlott, born in Kentucky (Lawson 1986:6). Nelson had five children at home (Jesse D., 21; Eliza, 19; James, 17; Frances, 16; and Harvey G., 9; all born in Kentucky) and $18,500 of property (Lawson 1986:6).

Another William Nelson was listed in the 1850 Montgomery County Census as a 52-year-old farmer born in Kentucky and married to 50-year-old Ann, born in Kentucky (Lawson 1986:20). Nelson had several children in his home (Nancy, 20; James F., 19; John B, 17; Susan Jane, 15; Selby F., 13; Emily T., 11; Cerilda H., 9; and Theodocia H., 7; all born in Kentucky) and $16,000 of property (Lawson 1986:20). In 1860, a William Nelson was listed in Montgomery (Jackson 1988a:761) and in Clark County in 1870 (Jackson 1988b:454). The Nelsons appear to be property owners in the vicinity of the millstone quarries.

Pigg Family

Anderson Pigg

Anderson Pigg is associated with the millstone quarries in the early 19th century. He was the son of William Pigg who moved to Lincoln County, Kentucky, about 1789 (Krauss 1982:76). Anderson was probably born in Pittsylvania County, Virginia, on July 12, 1772. He married Polly Perry in April 22, 1793 (Krauss 1982:76–77; Pigg Family 1985:7). Anderson Pigg was the nephew of Mary Ann (Pigg) Adams who married Elkanah Adams (Krauss 1982:78).

Krauss (1982:75) noted that the Pigg and Adams families started intermarrying when they lived in Pittsylvania County, Virginia. Adams (1981:1) stated, "The Adams family migrated to Kentucky about 1797 from Pittsylvania Co., Virginia. Census records, deeds, and other records from Clark and Madison counties, Kentucky and from Pittsylvania Co., Virginia suggest that several related families migrated together, and initially settled in Clark Co. in close proximity to one another. These families included Spencer and Sarah Corbin Adams, and possibly Elkanah and Harris Adams with their families, Sarah's father Rawley Corbin and his family, and members of the Pigg family."

In 1800, Anderson Pigg was listed in Lincoln County, Kentucky (Clift 1966:231). The U.S. Population Censuses for 1810 and 1820 listed Anderson Pigg in Clark County, Kentucky (Felldin and Inman 1981:223; Heritage Quest Online; Hubble 1992:138). In 1830, Anderson Pigg was listed in Estill County, Kentucky (Jackson, Teeples, and Schaefermeyer 1976:144). Anderson Pigg was listed in the 1840 U.S. Population Census for Clark County, Kentucky (Jackson and Teeples 1978c:174; Norris 1983:288).

The 1852 Tax Records for Powell County indicated that Anderson Pigg, who was white and over 21 years old, owned 50 acres of land (valued at $200) along the Beech Fork, and owned one horse (value of $40) for a total value of $240.

Anderson Pigg (July 12, 1772, to December 2, 1854) and his wife Polly (September 1, 1778, to February 2, 1846) are buried in Powell County, Kentucky (Martin 1999:99). It is uncertain whether Anderson Pigg was a millstone maker but he was definitely one of the early quarry owners.

Woodford Pigg

The 1852 Tax Records for Powell County indicated that Woodford Pigg owned 50 acres of land (valued at $200) along the Beech Fork. He had one white male over 21 years and owned one horse (value of $75), one mule ($50), and 13 cows ($130). He had a total value of $275. He also had two children between 6 and 18 years of age.

Woodford Pigg was listed in the 1850 Montgomery County Census as a 30-year-old farmer born in Kentucky and married to 24-year-old Elizabeth, born in Kentucky (Lawson 1986:48). No children were listed and he had $200 of property (Lawson 1986:48). The 1860 U.S. Population Census for Powell County listed Woodford Pigg as a 40-year-old man living at Stanton (Heritage Quest Online; Patrick 1981:3; Wonn 1981:4). He was married to 33-year-old Elizabeth with children William (14 years), Mary (12 years), Jinnett (9 years), John (6 years), Caroline (4 years), and Martha (2 years). He had $700 of real estate and $600 of personal property (Heritage Quest Online; Wonn 1981:4). Woodford Pigg was probably the son of Anderson Pigg and was likely a local landowner.

Risk Family

William Risk

William Risk was an early settler in Clark County, Kentucky. When he settled near Kidville in August of 1793, he noted that millstones were quarried prior to his arrival. He mentioned Peter DeWitt in his interview. William Risk was listed in the Clark County, Kentucky, tax lists for 1793 and 1799 (T. L. C. Genealogy 1990:9, 69). In 1800, William Risk was living in Clark County, Kentucky (Clift 1966:248). Risk was included in the U.S. population

Census for Clark County, Kentucky, in 1810 (Hubble 1992:140; Jackson and Teeples 1978a:695). He was also listed in the 1840 U.S. Population Census for Clark County, Kentucky (Jackson and Teeples 1978c:187; Norris 1983:276). The 1850 Clark County Census listed William Risk as a 76-year-old farmer born in Virginia and married to 56-year-old Polly Anderson Risk, born in Kentucky (Couey 1975:46). Thirty-two-year-old Isom Reynolds (his wife and 3 children) and William H. Anderson (14 years old) were living in Risk's household. He had $6,000 worth of property (Couey 1975:46). Risk was not involved in the millstone business but was familiar with the quarry and knew Peter DeWitt.

A transcribed version of William Risk's interview (John Shane, Draper MSS. 11CC 86) by Joseph R. Johnson (1965:6) contained the following note: "William Risk died on January 13, 1854. His wife Rachel Miller Risk died on October 1, 1837. On November 7, 1837, he married as his second wife Polly Anderson."

Ross Family

John A. J. Ross

In 1800, a John Ross was listed in Montgomery County, Kentucky (Clift 1966:254). During 1810, there were three John Rosses living in Estill County (Jackson and Teeples 1978a:672–673). There was a John Ross in the 1820 Census for Montgomery County, Kentucky (Felldin and Inman 1981:244). In 1830, John Ross was listed in Montgomery County, Kentucky (Jackson, Teeples, and Schaefermeyer 1976:157). John Ross appears to be a local property owner.

Smith Family

William Smith

William Smith was listed in the Clark County, Kentucky, tax lists for 1793, 1794, 1775, 1796, and 1799 (T. L. C. Genealogy 1990:9, 19, 34, 51, 70). In 1800, William Smith was listed in Clark County, Kentucky (Clift 1966:274). There was a William Smith listed in the 1820 population Census for Clark County, Kentucky (Heritage Quest Online). Forty years later, a William Smith was listed in the 1860 Clark County Census as a 54-year-old carpenter born in Kentucky and married to 36-year-old Sarah A., born in Kentucky (Norris 1981:1). Smith had six children living in his home (James W., 15; John C., 12; Susan, 10; Kiziah, 7; Sarah, 4; and Nancy, 1; all born in Kentucky) and $150 of property (Norris 1981:1). The 1870 Powell County Population Census listed a 33-year-old William M. Smith (Patrick 1988:10). Obviously, there were different William Smiths living in the area. They were probably just local land owners.

Spry Family

Cornelius Spry

Cornelius Spry gave a deposition in a lawsuit on behalf of Martin Johnson's son at the Office of Allan & Simpson in the town of Winchester on Saturday the 19th day of October 1822 (see Appendix E). He made millstones for Spencer Adams in 1819 and was a coworker of Martin Johnson during that period. Spry does not appear in the early tax records for Clark or Montgomery counties. He is not listed in the indexes for the Kentucky Population Census records for 1790, 1800, 1810, 1820, 1830, and 1840. Clark County marriages records indicate that Cornelius Spry married Lucy Wright on January 15, 1822 (Anonymous n.d.b:49). No other details were found concerning Spry. Apparently, Spry was not in Kentucky during census years and may have lived in the Commonwealth only briefly.

Stewart Family

Elijah Stewart

In 1800, Elijah Stewart was listed in Bourbon County, Kentucky (Clift 1966:282). An Eli Stewart was included in the 1840 Population Census for Montgomery County, Kentucky (Jackson and Teeples 1978c:212). Eli Stewart was listed in the 1850 Montgomery County Census as a 44-year-old farmer born in Kentucky and married to 43-year-old Sidney, born in Kentucky (Lawson 1986:7). Stewart had six children at home (Sophia, 17; Marchessa, 14; Elisha, 12; L. M., 10; William G., 6; and Amanda B., 4; all born in Kentucky) and $500 of property (Lawson 1986:57). Obviously, Elijah and Eli were not the same person but lived in the same area at different times. They were probably just local property owners.

James Stewart

On July 23, 1789, a James Stewart was listed in the Fayette County, Kentucky, tax list (Heinemann and Brumbaugh 1938:11). There was also a James Stewart living in Woodstock in Ulster County, New York, in 1790 and at Kingston (Ulster County), New York, in 1800 (Heritage Quest Online). James Stewart was listed in the Clark County, Kentucky, tax lists for 1793, 1794, 1795, 1796, 1797, and 1799 (T. L. C. Genealogy 1990:9, 19, 35, 52, 62, 70). In 1800, James Stewart was included in the Clark County, Kentucky, Census (Clift 1966:282). The U.S. Population Census for 1810 listed two James Stewarts in Clark County and one in Montgomery County, Kentucky (Heritage Quest Online; Jackson and Teeples 1978a:748). The U.S. Population Census for 1820 listed two James Stewarts in Clark County and one in Montgomery County, Kentucky (Heritage Quest Online). In 1830, two James Stewarts were listed in Montgomery County, Kentucky (Jackson, Teeples, and Schaefermeyer 1976:174). The 1840 Clark County Census listed James Stewart (Jackson and Teeples 1978c:212).

There were no James Stewarts in Ulster County, New York, in 1810 or 1820 (Felldin and Inman 1981:271; Heritage Quest Online).

James Stewart was listed in the 1799 tax records for Montgomery County, Kentucky, and owned four horses (Dunn 1996a:36). James Stewart married a woman named Ann in 1803 (Boyd 1961:99). It is unclear which James Stewart lived near the millstone quarries and if he had any involvement in the millstone industry.

Summers Family

Cornelius Summers

Cornelius Summers was listed in the Montgomery County, Kentucky, U.S. Population Census during 1810 (Dunn 1996b:2; Jackson and Teeples 1978a:760; Lawson 1985:12). The

U.S. Population Census for 1820 listed Cornelius Summers as living in Montgomery County, Kentucky (Heritage Quest Online). Cornelius Summers was married to Elizabeth Hadden by 1828 (Boyd 1961:101).

Information on the Summers Family in the vertical files at the Kentucky Historical Society (Anonymous n.d.a) provided the following details about Summers: "Cornelius Summers was born Jan. 5, 1790 and died May 24, 1845. This is the Reverend C. Summers, who was a circuit preacher. At Mt. Sterling, Kentucky, he married Elizabeth Hadden, who was born Feb. 18, 1791 and died Oct. 2, 1869. Both are buried in a cemetery near New Winchester, Indiana. They moved to Indiana in 1835."

On October 10, 1823, Summers bought the millstone quarry property from Spencer Adams. He later sued Spencer Adams in 1826 (Clark County, Kentucky 1826). Summers sold the quarry property to Anderson Pigg on August 27, 1835.

Treadway Family

John Treadway

John Treadway was the son of John Treadway and brother of Moses Treadway. John was born in 1769 and married Elizabeth Griffin (Krauss 1982:63). John Treadway was listed in Fayette County on February 22, 1790 (Heinemann 1976:95; Heinemann and Brumbaugh 1938: 11). He was also included in the Clark County, Kentucky, tax lists for 1793, 1794, 1795, and 1796 (T. L. C. Genealogy 1990:10, 20, 34, 53). John Treadway, Sr., and John Treadway, Jr., were both listed in the 1797 and 1799 tax records for Montgomery County, Kentucky (Dunn 1996a:25, 37). The 1800 tax records for Montgomery County, Kentucky, indicated that John Tradway [sic] owned 50 acres on Somerset and had two horses (Dunn 1996a:61). In 1800, two John Treadways were listed in Montgomery County, Kentucky (Clift 1966:298). John Treadway, Sr. and Jr., were both listed in the U.S. Population Census for Montgomery County during 1810 (Dunn 1996b:1; Jackson and Teeples 1978a:792; Lawson 1985:12). The U.S. Population Census for 1820 listed John Treadway and John Treadway, Sr., in Montgomery County, Kentucky (Felldin and Inman 1981:286; Heritage Quest Online; Lawson 1985:22). In 1830, John Treadway, Jr. and Sr., were listed in Montgomery County, Kentucky (Jackson, Teeples, and Schaefermeyer 1976:185). U.S. Population Censuses for Montgomery County during 1830 (Lawson 1985:42) and 1840 (Lawson 1985:58) also listed John Treadway. Either John Treadway, Sr. or Jr., may have been involved in the millstone industry since they were part of a millstone making family.

Another John Treadway was listed in the 1850 Montgomery County Census as a 50-year-old farmer born in Kentucky and married to 41-year-old Polly M., born in Kentucky (Lawson 1986:26). This Treadway had two children at home (Nancy A., 12; and Mary F., 10; both born in Kentucky) and $600 of property (Lawson 1986:26). This appears to be the John D. Treadway (April 13, 1800, to October 15, 1855) who was buried in a cemetery on the old Heaton Place in Montgomery County (Boyd n.d.).

Moses Treadway

Moses Treadway was a key person in the Powell County millstone industry. He was the son of John Treadway and the brother of John Treadway, Jr. Moses was born sometime between 1765 and 1775 and married Catherine DeWitt (Krauss 1982:63). Treadway (1951:44) suggested that Moses Treadway was the son of William Treadway of Baltimore County, Maryland, and first came to Kentucky in 1782. Further, Treadway (1951:45) stated that Moses Treadway married Mary Catherine DeWitt, daughter of Peter DeWitt, on July 9, 1794, with his brother John Treadway serving as a witness. He died sometime between 1820 and 1830 (Bowman 1963:7). Moses and Catherine had seven children; the oldest child was Peter DeWitt Treadway, who was born June 2, 1796, and died February 16, 1871 (Treadway 1951:45). Krauss (1982:63) stated that "all available records suggest that John and his family were the only Treadways living in Kentucky before 1800. They mostly lived on Lulbegrud Creek, a tributary of Red River, in Montgomery County."

Moses Treadway was listed in the Clark County, Kentucky, tax lists for 1795, 1796, 1797, and 1799 (T. L. C. Genealogy 1990:36, 53, 62, 70). In 1800, Moses Treadway was included in the Clark County, Kentucky, Census (Clift 1966:298). Treadway was listed in the U.S. Population Census, Clark County, Kentucky, during 1810 (Hubble 1992:143; Jackson and Teeples 1978a:792). The U.S. Population Census for 1820 listed Moses Treadway in Montgomery County, Kentucky (Felldin and Inman 1981:286; Heritage Quest Online; Lawson 1985:22). Treadway was a millstone maker who married into the DeWitt family that also made millstones.

The 1852 Tax Records for Powell County indicated that Moses H. Treadway owned 90 acres of land (valued at $500) along the Lulbegrud Creek and had one white male over 21 years. He owned two horses and mares (value of $100) and three cows ($20), and had a total value of $1,320. This may have been a son of Moses Treadway.

Peter Treadway

Peter Treadway was the oldest child of Moses Treadway and Catherine DeWitt Treadway (Bowman 1963:7; Treadway 1951:45). He was born June 2, 1796, and died February 16, 1871 (Treadway 1951:45). One source suggests that Treadway married Margaret Evans (Bowman 1963:7). Treadway (1951:45) indicated that Peter married Margaret Euins (May 4, 1806–August 12, 1874) and that the union produced 11 children including a son, Moses X. Treadway (April 8, 1826–January 7, 1899). In 1830, Peter Treadway was listed in Montgomery County, Kentucky (Jackson, Teeples, and Schaefermeyer 1976:185). The U.S. Population Censuses for Montgomery County during 1830 (Lawson 1985:42) and 1840 (Lawson 1985:58) listed Peter Treadway.

The 1852 Tax Records for Powell County indicated that Peter Treadway owned 100 acres of land (valued at $100) along Red River and one horse (value of $50) for a total value of $150. He also had three children between 6 and 18 years of age.

Peter Tredaway [sic] was listed in the 1850 Clark County Census as a 54-year-old stone cutter born in Kentucky and married to 44-year-old Margaret, born in Kentucky (Couey 1975:36). Tredaway had eight children at home (Simeon, 22; Margaret, 19; John, 17; Sarah A., 16; Albert G., 14; Louisa, 10; Bluford, 7; and Charles D., 2; all born in Kentucky) and $500 of property (Couey 1975:36). The U.S. Population Census for 1860 listed a 61-year-old Peter Treadway in Estill County, Kentucky, who was married to 52-year-old Margaret (Heritage Quest Online). Since Peter Treadway was the son of millstone maker Moses Treadway and was listed in the 1850 Population Census as a stone cutter, he probably worked at the Red River millstone quarry.

Ware Family

Achilles "Killis" Ware

The 1910 Powell County Census listed 24-year-old Killis Ware (Hensley 1986:6). He was married to 27-year-old Dora with daughter Ethel (2 years) and son Boone (1 year) (Hensley 1986:6). Killis Ware (February 17, 1887 to November 14, 1933) was buried in the Virden Cemetery in Powell County (Martin 1999:133). He was probably a local land owner.

Dillard P. Ware

In the 1860 Clark County Census, Dillard Ware was a 3-year-old boy living with his parents, Robert (37 years old) and Emily A. (34 years old) Ware (Norris 1981:98–99). The 1900 Powell County Census listed Dillard P. Ware as a minister and his wife as a school teacher. The 42-year-old Ware was married to 35-year-old Pattie A. and had a daughter named Cora (17 years) (Heritage Quest Online; Morton 1995:15). A Rev. D. P. Ware (July 23, 1857 to December 15, 1904) and his wife Sallie Ware (July 1, 1861 to July 2, 1898) were buried in a Winchester Cemetery in Clark County (Owen and Couey 1983:116). Undoubtedly, Ware was just a property owner in the vicinity of the quarries.

Thomas B. Ware

The 1900 Powell County Census listed Thomas B. Ware as the sheriff. The 49-year-old Ware was married to 49-year-old Martha with children Nannie (26 years), James H. (21 years), Charles (19 years), Lizzie B. (17 years), Killis (14), and Mary (11) (Heritage Quest Online; Morton 1995:15). The 1910 Powell County Census listed Thomas B. Ware as a 59-year-old man married to 59-year-old Martha with children Robert (38 years), Lillie (26 years), Mary (20 years), and Grace (17 years) (Hensley 1986:5). Thomas B. Ware was also listed in the 1920 Powell County Census as a 68-year-old man married to 69-year-old Martha with children Mary (31 years) and Lillie (38 years) still at home (Morton 1993:3). Thomas B. Ware (July 17, 1858, to March 23, 1913) is buried in the Vaughns Mill Cemetery in Powell County (Martin 1999:133). Ware was a local land owner.

West Family

Harrison West

In the 1900 Montgomery County Census, Harrison West was listed as a 33-year-old farmer. He was married to 26-year-old Donna with children Annie (6 years), Strocker (4 years), and Vernon (2 years) (Heritage Quest Online). Apparently, he owned land near the millstone quarries after they were abandoned.

Stone Cutters

The 1850 Clark County Census listed six stone cutters (Couey 1975:36, 100, 102, 120). Several individuals were listed as stone masons and stone cutters, which are different occupations. *Webster's New Twentieth Century Dictionary of the English Language* (McKechinie 1978:1794) defined stonemason as "a person who cuts stone to shape and use it in making walls, buildings, etc." Stonecutter, on the other hand, is "one whose occupation is cutting and dressing stones; also a machine for trimming stones" (McKechinie 1978:1793). Since stone cutters are usually associated with quarries, these individuals are potentially workers at the millstone quarries. Table 1 includes these individuals, where they were born, and ages.

The 1860 Clark County Census also listed four stone cutters (Norris 1981:23, 30, 68, 70). These stone cutters are potentially workers at the millstone quarries. Table 2 lists these individuals, where they were born, and ages.

The men listed in Tables 1 and 2 may have been employees of the Red River millstone quarries. Their occupations as stone cutters indicate that they worked at a quarry. Additional research will be required before they can be firmly connected to the millstone quarries. In the meantime, they remain prime candidates as millstone makers.

Table 1. Stone Cutters Listed in the 1850 Clark County, Kentucky, Population Census

Name	Age	Birth Place
Peter Treadway	54 years	Kentucky
John Deacon	33 years	Pennsylvania
John A. Brink	25 years	Kentucky
Joseph Deacon	37 years	Pennsylvania
James Connell	31 years	Ireland
David Shanks	38 years	Kentucky

Table 2. Stone Cutters Listed in the 1860 Clark County, Kentucky, Population Census

Name	Age	Birth Place
Joseph Deacon	47 years	Pennsylvania
John Coons	40 years	Kentucky
Benjamin Danley	26 years	Kentucky
Greenberry B. Parrish	36 years	Kentucky

6

Archaeological Investigations at the Powell County Quarries

This chapter describes the archaeological remains documented at the McGuire, Baker, Toler, Ware, Ewen, and Pilot Knob millstone quarries. The first part of each discussion describes the physical setting of the quarry. Subsequent paragraphs describe the millstones, drilled boulders, quarry excavations, and other remains in detail. These discussions are supplemented with photographs and tables. To reduce the potential for vandalism and theft of millstones, no precise location maps or specific quarry maps are included in this book. However, an interpretive trail has been developed at the Pilot Knob quarry that is located at the Pilot Knob State Nature Preserve. A booklet which describes the quarry and millstone manufacture is available at the park for self guided tours of the quarry (Hockensmith 1993a, 1994a). The McGuire quarry was acquired by Powell County several years after this fieldwork (Meadows 2002). Initially, trails were planned through the quarries but it was later decided to keep these quarries closed to the public.

McGuire Quarry

The McGuire Millstone Quarry (15Po305) is located on Rotten Point knob (Figure 4). Field work was conducted on March 17 and April 1–2, 1987; March 8, 9, 15, 22, 1988; March 2 and April 12, 19–20, 1990. Quarrying remains occur in a crescent shaped area approximately 600 m long that varies in width from 40 to 105 m. The quarry can be subdivided into three distinct areas referred to as the Upper, Middle, and Lower quarries. The Upper Quarry is situated near the knob crest at about 1,360 feet above Mean Sea Level (MSL). This area is approximately 130 m north-south and 80 m east-west (1.04 ha). A large bench (46 m north-south and 10 m east-west) has been excavated on the western edge of the knob crest (Figure 5). The bench contains a number of conglomerate boulders (some with drill holes) and some unfinished millstones on the slope above it. Immediately east (upslope) of the bench is an oval pit (12 × 8 m), ca. 2–3 m deep. A few meters to the southeast is the edge of a large crescent shaped pit (Figure 6). The pit (2–4 m deep) curves 46 m to the southeast and varies from 2 to 5 m in width. On the southeast end, the pit opens into a narrow (up to 6 m wide) bench about 27 m long with a rear wall about 3 m high. Six incomplete millstones and 16 boulders with drill holes were recorded in this area. This area also contains numerous conglomerate fragments. In the Upper Quarry, the millstone makers removed the overburden to expose the in situ layers of conglomerate.

The Middle Quarry is immediately down slope from the Upper Quarry (ca. 1100–1200 feet above MSL). This portion of the quarry is on slopes that are quite steep in places. Quarry remains are scattered over an area about 40 m east-west and 155 m north-south (0.62 ha). This area contains two narrow benches and an irregular pit. The largest pit measures 15 × 3 m and is 1 to 1.5 m deep. Six millstones, 15 boulders with drill holes, and numerous conglomerate fragments are scattered throughout this area. The majority of the quarrying remains are located in the northernmost 75 m of this area while the southern portion contains only sparse remains.

The Lower Quarry is southeast of the Middle Quarry and extends across four narrow ridges. Additional remains are located on the steep side slopes and within the streambeds between the ridges. This portion of the quarry ranges from about 800 to 1,200 feet above MSL. The overall dimensions of the Lower Quarry are about 315 m north-south and 105 m east-west (3.3 ha). For recording purposes, the ridges were designated A, B, C, and D. Likewise, the streams between the ridges were assigned the letter designations A, B, C, D, and E. The ridges are roughly parallel to one another and are oriented north-south.

Ridge A contains a sparse scatter of conglomerate boulders along its southern end (Figure 7). Quarry remains include four millstones, one boulder with drill holes, and conglomerate fragments. Adjacent to Ridge A is Ridge B which is much longer. The crest and slopes of Ridge B contains a very sparse scatter of conglomerate boulders, shaping debris, three oval quarry pits (2 × 2 × 0.5 m; 2 × 2 × 0.5 m; 5 × 5 × 1.5 m) (Figure 8), three millstones, and five boulders with drill holes. Stream A along its western border contains four millstones and five boulders with drill holes. Located along the eastern edge of Ridge B, Stream C contains two millstones and three boulders with drill holes. Ridge C is very narrow and only contains

Figure 4. View of Rotten Point Knob looking north from the access road. Photograph taken on March 22, 1988.

Figure 5. Quarry bench at the Upper McGuire Millstone Quarry. Photograph (facing north) taken on April 2, 1987.

Figure 6. Quarry pit at the Upper McGuire Millstone Quarry. Photograph (facing east) taken on April 2, 1987.

Figure 7. East end of Ridge A at the Lower McGuire Millstone Quarry. Photograph (facing west) taken on April 12, 1990.

Figure 8. Small pit with boulders at the end of Ridge B at the Lower McGuire Millstone Quarry. Photograph (facing south) taken on April 20, 1990.

three millstones and some boulders on its crest and slopes. The segment of Stream D between Ridges C and D contains five millstones and seven boulders with drill holes. Further east on Ridge D, four millstones and 14 boulders with drill holes were documented. The ridge also contains a large bench (20 m east-west and 80 m north-south) which has an oval quarry pit (4 m by 5 m and 1 m deep). Another oval quarry pit (3 m by 4 m and 1.5 m deep) is located on the crest of Ridge D.

During the fieldwork, 38 millstones and 66 boulders with drill holes were documented for the quarry as a whole. The conglomerate used for millstones is light gray to light tan sandstone containing rounded quartz pebbles (mostly white, some yellow and brown). Most of the pebbles are under 2 cm in diameter with some up to 4 cm across.

Millstones

Thirty-eight millstones were documented at the McGuire Quarry. Table 3 provides a summary of the diameter, thickness, and central hole diameter for each millstone from this quarry. Millstones 3–10 are located in the Upper Quarry; Millstones 1, 2, 11, and 12 are within the Middle Quarry, and Millstones 13–38 are situated in the Lower Quarry. The following discussion describes these millstones and offers possible reasons on why they were abandoned.

Millstone #1 is a nearly complete specimen (Figure 9). This millstone is 103 cm in diameter, 35 cm thick, and has an eye 17 cm in diameter that goes through the specimen. It appears to have been abandoned due to a depression (20 × 25 cm and 4 cm deep) on the upper surface and edge damage. The depression may be the result of a poorly cemented large fossil or other inclusions that separated from the stone during manufacture. The edge damage may be the consequence of a misplaced blow during shaping or a flaw in the stone. Regardless of the cause, the stone was rejected because the entire surface would had to have been worked down 4 cm and the diameter reduced beyond the edge damage.

Figure 9. Millstone # 1 at the Upper McGuire Millstone Quarry. Photograph (facing west) taken on April 1, 1987.

Table 3. Millstone Measurements at the McGuire Quarry

Specimen No.	Diameter/Size (cm)	Thickness (cm)	Eye Diameter (cm)
01	103	35	17
02	90	35	17
03	90	22* (60)	18
04	105	30	17
05	110	20	NA
06	93 × 113	40	NA
07	115	36	NA
08	128	52	NA
09	113 × 123	52	NA
10	103 × 125	25	NA
11	105	42	NA
12	110	42	NA

Specimen No.	Diameter/Size (cm)	Thickness (cm)	Eye Diameter (cm)
13	102	21	21
14	102	27	17.5
15	122	14* (55)	20
16	110 × 145	45+	NA
17	115 × 135	60	NA
18	90 × 105	30	NA
19	90 × 105	40	NA
20	104 × 122	30	NA
21	80 × 90	25	NA
22	150 × 208	50	NA
23	64 × 92	20	NA
24	90 × 92	30	NA
25	35 × 120	40	NA
26	110	35	NA
27	106	24	22
28	85 × 100	32	NA
29	97	46	NA
30	100 × 110	22	NA
31	88	14* (32)	18
32	100 × 110	26	NA
33	110 × 124	20	NA
34	115 × 130	60	NA
35	110 × 120	40+	NA
36	110 × 115	32	NA
37	125	50	NA
38	100 × 115	70	NA

*Thickness of shaped portion.
+Bottom buried and maximum thickness unavailable.

Millstone # 2 is half of a nearly complete specimen (Figure 10). The stone is 90 cm in diameter, 35 cm thick, and has an eye 17 cm in diameter. The stone appears to have split into halves when the eye was being cut. The eye was cut 22 cm deep. Millstone maker Houston Surface noted that cutting the eye was a critical time when the stone could fall apart (Hockensmith and Coy 1999:30). The millstone maker would cut the eye halfway through the stone and finish the hole from the other side. This specimen apparently split apart before the eye could be completed.

Millstone # 3 is a nearly complete specimen still attached to the parent boulder (Figure 11). The stone is 90 cm in diam-

Top right: Figure 10. Millstone # 2 at the Upper McGuire Millstone Quarry. Photograph (facing west) taken on April 1, 1987.

Right: Figure 11. Millstone # 3 at the Upper McGuire Millstone Quarry. Photograph (facing east) taken on April 1, 1987.

eter, 60+ cm thick, and has an eye 18 cm in diameter. The sides of the stone were shaped to 22 cm below the top. The eye was partially cut to a depth of 9 to 12 cm. Two areas of edge damage are present. The larger fracture resulted in a depression 3–4 cm deep in the upper surface. The stone was rejected because the entire surface would had to have been worked down 4 cm to level the top where the edge damage occurred.

Millstone # 4 is a nearly complete specimen (Figure 12). The stone is 105 cm in diameter, 26–30 cm thick, and has an eye 17 cm in diameter that goes through it. The face of the stone is slightly rough with shallow grooves on this surface. These grooves may be drill hole scars from splitting the millstone from the parent boulder. The six grooves are in a radial pattern (farther apart at the outer edge and closer together towards the eye). They range between 29 and 34 cm apart near the outer edge of the stone. Three of the grooves have a spacing of 33 cm. Wedges and feathers were probably used to separate the millstone from the boulder. Two areas of edge damage are present. One irregular area, measuring about 25 × 25 cm, ranges in depth between 3 and 8 cm. It would not have been feasible to lower the entire face of the stone by 8 cm. A tree is growing through the eye of the millstone.

Figure 12. Millstone # 4 at the Upper McGuire Millstone Quarry. Note the tree growing through the eye of the millstone. Photograph (facing east) taken on April 2, 1987.

Millstone # 5 is a rough preform. The stone has a diameter of 110 cm in diameter and is 20 cm thick. The upper face was flattened. It appears that the stone broke in an irregular manner during shaping which would have made it undesirable to reduce the diameter further.

Millstone # 6 is a roughly shaped preform. The stone measures 93 × 113 cm and is 40 cm thick. The stone appears to have been rejected because of irregular breaks during the shaping process. A single vertical drill hole scar is present along one of the broken edges. The drill hole is 3 cm in diameter and 10 cm deep. The drill hole may have been employed to remove excess stone from beyond the projected circumference.

Millstone # 7 is a slab that has been rounded on one end but is irregular on the other end. The stone is about 115 cm in diameter and is 35+ cm thick. The upper face is relatively flat. Four vertical drill holes are present on the irregular end, which shows how the slab was separated from the larger parent rock. The drill holes are 3 cm in diameter and range in depth from 4 and 25 cm (4, 6, 12, and 25 cm apart). Drill holes 1 and 2 are 40 cm apart while drill holes 3 and 4 are 20 cm apart. The rounded portion of the stone has three shallow (2 cm deep) depressions on the upper face: 18 × 60 cm; 12 × 16 cm; and 11 × 28 cm. The stone was rejected because the entire surface would had to have been worked down at least 2 cm and the diameter reduced beyond the edge damage.

Millstone # 8 is an advanced stage preform. The stone measures 128 cm in diameter and is 52 cm thick. The specimen has been completely rounded but the upper face has a somewhat irregular surface. This exposed face rages from 6 to 7 cm from being flat in areas. Two areas have edge damage; the largest area is about 45 × 45 cm. Two shallow depressions are present near the center of the face. The stone was rejected because the entire surface would had to have been worked down.

Millstone # 9 is an advanced stage preform located in a small pit (Figure 13). The stone measures 113 × 123 cm and is 52 cm thick. The specimen is almost perfectly level on the upper face. About half the stone is well rounded and the other half is very roughly rounded. During the shaping process, the stone broke in an irregular manner which undercut the bottom face. The stone was rejected because of this flaw. On one side of the stone, five vertical tool marks are present. These tool marks are poorly defined and are not drill holes. They range in length between 12 and 20 cm (12, 15, 18, 20, and 20 cm long) and are spaced between 11 and 21 cm (11, 14, 16, and 21 cm apart). It is not known what type of tool produced these marks.

Figure 13. Millstone # 9 at the Upper McGuire Millstone Quarry. Photograph (facing northeast) taken on March 8, 1988.

Millstone # 10 is a roughly shaped slab that has been rounded at one end and is oval at the other end. The stone measures about 103 × 125 cm and is 25 cm thick. The upper face is relatively flat except for one projecting edge which about 5 cm higher. Problems were encountered while shaping the stone resulting in irregular breaks around the circumference. Also, a 56 cm long crack is present along the round edge of the millstone.

Millstone # 11 is a large quarried slab that has been rounded about two-thirds the way around the circumference. The stone measures 100 × 156 cm and is 42 cm thick. The specimen has been completely rounded but the upper face has a somewhat irregular surface. The rounded end would have produced a millstone about 105 cm in diameter. The rough side has three drill holes. Two drill holes are across the upper face of the millstone and apparently were used to thin the slab. These holes are 3 cm and 4 cm in diameter, respectively, and are 27 cm apart. They are 14 and 16 cm deep. The vertical drill hole is on the opposite side from the horizontal holes. Apparently, this hole was intended to help shape the circumference of the millstone. It is 3 cm in diameter and 8 cm deep. The millstone makers may have experienced problems in shaping the stone.

Millstone # 12 is a roughly shaped preform. The stone is 110 cm in diameter and is 42 cm thick. It is roughly rounded and the upper surface is flattened. It contains about five scars from horizontal drill holes or tool marks on the upper surface. Since these tool scars always extend from the edges of the stone, they were probably connected with the leveling of the upper surface. The stone appears to have been rejected because of irregular breaks during the shaping process, and it also has a fracture across one edge.

Millstone # 13 is a nearly perfect specimen (Figure 14). The specimen is 102 cm in diameter, 21 cm thick, and has an eye 21 cm in diameter that goes through the stone. There

Figure 14. Millstone # 13 at the Lower McGuire Millstone Quarry. Photograph taken on March 17, 1988.

are no visible flaws on the exposed side of the stone. It may have an irregular back but this cannot be determined without turning the stone over.

Millstone # 14 is a nearly perfect specimen (Figure 15). The stone is 102 cm in diameter, 27 cm thick, and has an eye 17.5 cm in diameter that goes through the stone. There is some minor edge damage on the upper surface. It is uncertain whether this damage occurred during shaping or afterwards. There could also be additional damage on the reverse side.

Figure 15. Millstone # 14 at the Lower McGuire Millstone Quarry. Photograph taken on March 17, 1988.

Millstone # 15 is an advanced stage preform (Figure 16). The stone measures 122 cm in diameter, is 55 cm thick, and the eye is 20 cm in diameter. This is an extremely interesting specimen since the unfinished millstone is still attached to the parent boulder. The upper face of the stone was within 3 cm of being level. There are two shallow depressions 8 × 18 cm and 18 × 25 cm on opposing edges of the upper surface. These depressions are in a straight alignment with the eye. Both depressions are 2 cm deep and may be the beginning stage of a leveling trough. The upper 10–14 cm of the sides has been rounded. The eye was cut to a depth of 18 cm. The only visible flaw that would have caused the stone to be rejected was an area of

Figure 16. Photograph of Millstone # 15 at the Lower McGuire Millstone Quarry. Photograph (facing north) taken on April 12, 1990.

edge damage 8 × 14 cm and 5 cm deep. The attached boulder below the millstone extends from 14 to 49 cm beyond the millstone and is 25 to 30 cm thick.

Millstone # 16 is a roughly shaped preform. The stone measures 110 × 145 cm and is 45+ cm thick. This oval shaped boulder has been rounded on one end. The upper surface is irregular, varying as much as 15 cm. The sides are sloping on the rounded part and irregular elsewhere. The stone appears to have been rejected because of irregular breaks during the shaping process.

Millstone # 17 is a large rectangular quarried slab that has been slightly rounded at each end. The stone measures 115 × 135 cm and is 60 cm thick. The upper surface is relatively flat but varies 2–3 cm. The sides are very irregular. This stone appears to have been rejected because of irregular breaks during the shaping process, cracks, and an undesirable lense of chert (35 cm long and 7 cm thick).

Millstone # 18 is a roughly shaped preform. The stone is 90 × 105 cm and is 30 cm thick. It is roughly rounded and the upper surface is irregular (varies 5–6 cm). The sides are irregular as well. The stone appears to have been rejected because of irregular breaks during the shaping process that undercut the base thus making it unsuitable for further shaping.

Millstone # 19 is a rough preform that is pear-shaped (Figure 17). The stone is 90 × 105 cm and is 40 cm thick. The upper surface is very irregular (varies 15–20 cm). One side is ver-

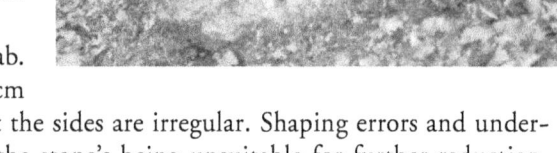

Figure 17. Millstone # 19 at the Lower McGuire Millstone Quarry. Note the leveling cross on the upper surface. Photograph (facing east) taken on April 20, 1990.

tical but is irregular around the remainder of the circumference. Irregular breaks during the shaping process made the stone too small for further shaping.

Millstone # 20 is an oval shaped slab. It measures about 104 × 122 cm and is 30 cm thick. The upper face is relatively flat but the sides are irregular. Shaping errors and undercutting on the southern side resulted in the stone's being unsuitable for further reduction. Another problem was that the pebbles were not sufficiently dense on the northern end.

Millstone # 21 is a roughly oval preform. The stone is 80 × 90 cm and is 25 cm thick. The upper surface is very irregular (varies 5–10 cm) with an oval projection and a larger depression. The sides are vertical but irregular. The stone was rejected due to shaping errors and undercutting.

Millstone # 22 is a large oval slab representing an early stage preform. The stone measures 150 × 208 cm and is 50 cm thick. The upper surface is relatively flat except for the northern end, which is 10 cm higher but varies 2–3 cm vertically. The sides are roughly vertical. The stone appears to have been rejected because of irregular breaks during the shaping process and minor undercutting.

Millstone # 23 is a small oval slab. It measures 64 × 92 cm and is 20 cm thick. The upper surface is relatively flat but varies 2–4 cm in elevation. The lower side is irregular but the remainder of the stone is buried. Shaping errors and undercutting on the exposed side were probably responsible for the stone being rejected.

Millstone # 24 is a roughly rectangular slab that has been worked. This early stage preform measures 90 × 92 cm and is 30 cm thick. The upper surface has been flattened but varies 1–5 cm. The sides are roughly vertical. Rejection was due to shaping errors. Also, a tree fossil in the upper surface made the stone undesirable. Four horizontal tool marks are present on the upper surface of the stone. These tool marks are poorly defined and are not drill holes. They range in length between 14 and 25 cm (14, 18, 24, and 25 cm long) and are spaced between 14 and 25 cm (14, 18, and 25 cm apart). The tool marks are 5 and 10 cm from the edges of the stone, respectively. They are ca. 2 cm wide and 1–2 cm deep. It is not known what tool type produced these marks. However, they may be chisel marks associated with leveling the top of the stone.

Millstone # 25 is an oval slab representing an early stage preform. The stone measures 85 × 120 cm and is 40 cm thick. The upper surface is slightly irregular and varies vertically 3–7 cm. The sides are roughly vertical. The southwest end of the stone was shattered during the shaping process and the stone was also undercut in two locations.

Millstone # 26 is an oval shaped preform (Figure 18). The specimen was rounded about two-thirds around the circumference.

Figure 18. Photograph of Millstone # 26 at the Lower McGuire Millstone Quarry. Photograph (facing west) taken on April 20, 1990.

The stone measures 110 × 110 cm and is 35 cm thick. The upper surface is relatively flat except for two areas. The sides are vertical. The stone was rejected due to shaping errors including a major break (37 cm long and 13 cm wide) along one edge. An unusual feature of this stone is a wedge shaped piece taken out of one side. One cut was 35 cm long and the other cut was 27 cm long. The depression was 4 cm deep and chisel marks (2 cm wide) are visible inside the cut. This may be the beginning of a leveling cross cut. However, the actual function is uncertain.

Millstone # 27 is a nearly complete specimen (Figure 19). The stone is 106 cm in diameter, 24 cm thick, and has an eye 22 cm in diameter. This specimen is located in a stream bank and about 40 percent of the millstone is below water level. The upper surface is flat and the edges are vertical. One edge has been chipped off (23 cm long, 10 cm wide and 1–9 cm deep) and other flaws may be present on the buried portion of the stone. Also, the eye goes through the stone but does not appear to be properly centered.

Figure 19. Millstone # 27 at the Lower McGuire Millstone Quarry in edge of Stream A at Ridge B. Photograph (facing north) taken on April 19, 1990.

Millstone # 28 is an oval slab representing an early stage preform. The stone measures 85 × 100 cm and is 32 cm thick. The specimen is well rounded about two-thirds around the circumference. The upper surface is flattened but varies vertically 4–5 cm. The sides are vertical but irregular. The southwest end of the stone was shattered during the shaping process and the stone was also undercut in two locations.

Millstone # 29 is a roughly rounded specimen. This stone is 97 cm in diameter and is 46 cm thick. The upper surface has been flattened but varies 4–5 cm. The sides are roughly vertical. Rejection was due to the northwest edge of the stone breaking off during the shaping process. The broken area is 55 cm long, up to 24 cm wide, and 15–22 cm deep. Also, a tree fossil on the upper surface made the stone undesirable. There are also two large holes where inclusions have eroded from the upper surface. Six horizontal tool marks are present on the upper surface of the stone. These tool marks are composed of two sets of three on opposing sides of the specimen. On the east side, they range in length between 15 and 22 cm (15, 17, and 22 cm) and are spaced between 5 and 8 cm apart. On the west side, they range in length between 11 and 17 cm (11, 12, and 16 cm) and are spaced between 5 and 17 cm apart. These tool marks may be chisel marks associated with leveling the top of the stone.

Millstone # 30 is an oval shaped preform that was rounded. The stone measures 100 × 110 cm and is 22 cm thick. The upper surface is flattened but varies 4–5 cm vertically. The sides are vertical but still irregular. The stone was rejected due to shaping errors including a large depression 25 × 23 cm and 4 cm deep. This specimen is located in a stream bank and the base of the millstone is below water level.

Millstone # 31 is a nearly complete specimen still attached to the parent boulder (Figure 20). The stone is 88 cm in diameter, 32 cm thick, and has an eye 18 cm in diameter. The stone was perfectly rounded 12–17 cm down from the top. The eye was partially cut to a depth of 8 cm. During the cutting of the eye, the stone fractured from the eye to the edge of the millstone. The resulting damage from this fracture was 34 cm long, 34 cm wide at the edge

Figure 20. Millstone # 31 at the Lower McGuire Millstone Quarry. Photograph (facing northeast) taken on April 19, 1990.

of the stone, and 1–7 cm deep. Two other areas of edge damage, 12 and 22 cm, were also present. The parent boulder is still attached to the bottom of the millstone.

Millstone # 32 is an oval shaped preform that was rounded. The stone measures 100 × 110 cm and is 26 cm thick. The upper surface is relatively flat but the sides are irregular. The stone was rejected due to shaping errors, including one area 40 cm long and 14 cm wide that broke along the edge.

Millstone # 33 is an oval shaped preform. The specimen was rounded about halfway around the circumference. The stone measures 110 × 124 cm and is 20 cm thick. The upper surface varies 1–4 cm vertically while the sides are irregular. The stone was rejected due to shaping errors. The upper surface of the millstone and the exposed side contains tool marks. Three linear grooves are present on the top of the south half of the millstone. The main groove (oriented east-west) is 65 cm long, 2–3 cm wide, and 1–3 cm deep. About 12 cm east of this groove is another groove at a slightly different angle (towards the southeast). The second groove is 15 cm long, 5 cm wide, and 2–3 cm deep. A third groove is 6 cm south of the main groove and is also parallel to it. This groove is 15 cm long, 1 cm wide, and 1 cm deep. Two vertical grooves (their size was not recorded) are on the south side of the stone. They are 18 cm apart while one is 22 cm from the west edge and the other is 15 cm from the east edge. These tool marks may be chisel marks associated with leveling the top of the stone and straightening the side.

Millstone # 34 is a roughly rectangular shaped boulder that has been rounded. This early stage preform measures 115 × 130 cm and is 60 cm thick. The upper surface has been flattened but varies 4–5 cm. The sides are nearly vertical. Rejection was due to shaping errors. On the west edge there is an irregular break 60 cm long, 20 cm wide, and 10 cm deep. A second break on the north end is 26 cm long, 15 cm wide, and 10 cm deep. Further, there is a void in the side (35 cm long, 15 cm wide, and 15 cm deep) where a fossil or other inclusion fell out.

Millstone # 35 is a roughly rounded specimen. This stone measures 110 × 120 cm and is 40+ cm thick. The upper surface is very irregular. About two-thirds of the boulder has been rounded. Rejection was due to an irregular break on the south edge (25 cm long and 10 cm wide). It was not feasible to further reduce the millstone blank.

Millstone # 36 is a roughly rectangular boulder that has been rounded on the corners (Figure 21). This early stage preform measures 110 × 115 cm and is 32 cm thick. The upper surface has been flattened but varies vertically 1–3 cm. The sides are irregular. The millstone was abandoned due to shaping errors and undercutting.

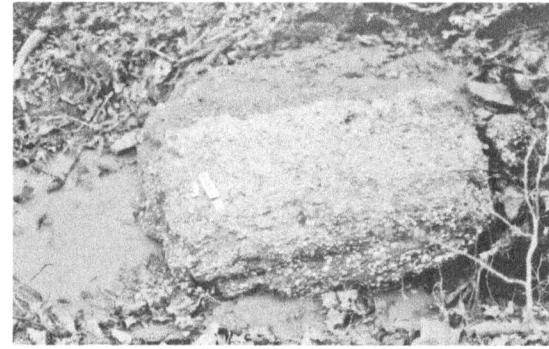

Figure 21. Millstone # 36 in Stream D at the Lower McGuire Millstone Quarry. Photograph (facing south) taken on April 19, 1990.

Millstone # 37 is a roughly rounded specimen (Figure 22). This stone is 125 cm in diameter and is 50 cm thick. The upper surface is very level (within 1–2 cm). About three-fourths of the boulder has been rounded. Shaping errors on the south side of the stone and a hole (30 cm long, 8 cm wide, and 8–17 cm deep) from a missing inclusion rendered the stone unusable. Four tool marks are present on the northern edge of the upper surface. The three horizontal marks on the northwest corner are 10 cm and 17 cm apart. On the east side of the specimen are three diagonal tool marks. They range in length from 20 to 32 cm (20, 20, and 30 cm long) and are spaced between 5 and 20 cm apart. These tool marks may be chisel marks associated with leveling the top of the stone and straightening the side.

Figure 22. Millstone # 37 at the Lower McGuire Millstone Quarry. Note vertical tool marks on the side. Photograph (facing west) taken on April 20, 1990.

Millstone # 38 is a roughly rounded specimen. This stone measures 100 × 115 cm and is 70 cm thick. The upper surface is relatively flat except for the edges where it varies up to 10 cm. The sides are very irregular. The base of the specimens flares outward. Rejection was due to shaping errors, a large crack, and undercutting on the north and south ends.

The 38 millstones from the McGuire Millstone Quarry represent several stages in the manufacturing sequence. These stages include nearly complete millstones (n=8), advanced preforms (n=3), oval or rounded preforms (n=23), and rectangular preforms (n=4). They range in diameter from 80 cm to 1.50 m but most rejected millstones cluster between 85 cm and 1.15 m. In terms of thickness, they vary from 20 cm to 70 cm with most specimens clustering between 20 cm and 40 cm. Millstones were rejected for a variety of reasons including irregular breaks (n=27), undercutting (n=9), edge damage (n=7), surface depressions (n=7), cracks (n=4), splitting apart (n=1), inclusions (n=4), and unknown (n=1). Sometimes two or three flaws occurred on the same specimen.

Other noteworthy features of some McGuire millstones include eyes and leveling crosses. No leveling crosses were encountered on the rounded millstones. However, Boulder # 21, a possible millstone preform, has a partial (T-shaped) leveling cross on the upper surface. The cross ranges in width from 20 to 24 cm and is 2 cm deep. Nine millstones (Numbers 1, 2, 3, 4, 13, 14, 15, 27, and 31) contained eyes. Five of the eyes were cut all the way through the millstones while the remaining four eyes were 8 to 22 cm deep (8, 9–12, 18, and 22 cm deep). The diameters of the eyes ranged from 17 to 22 cm with six examples between 17 and 18 cm (17, 17, 17, 17.5, 18, 18 cm in diameter) and three examples between 20 and 22 cm (20, 21, 22 cm in diameter).

Manufacturing evidence consists of linear tool marks and drill holes. Eight millstones (Numbers 4, 9, 12, 24, 26, 29, 33, and 37) had linear tool marks and three millstones (Numbers 6, 7, and 11) had drill holes. The horizontal linear tool marks (n=27) ranged in number from three to six and one specimen had a radial pattern (Millstone 4). Available size ranges for the horizontal marks were 14 to 65 cm long (most 14–25 cm), 1 to 5 cm wide (most 2 cm), 1 to 4 cm deep, and 5–25 cm apart (most 5–17 cm). Vertical tool marks (n=8) occurred in groups of three to five and ranged in length between 12 and 32 cm and 5 to 21 cm apart. The eight drill holes were 3 cm in diameter with one exception (4 cm), ranged in depth between 4 and 25 cm (most 10–16 cm), and had hole spacing ranging from 20 to 40 cm.

Two isolated millstones were present in the vicinity of the McGuire Quarry. Millstone # 4 in the collection of the Red River Historical Society was discovered near the Lower McGuire Quarry at the mouth of a small valley (see Appendix G). This specimen is slightly less than half of a complete millstone. It is 43 cm in diameter and 22 cm thick. The eye is 10 cm in diameter and was cut to a depth of 4 cm. It appears to have split apart when the eye was cut. Also, an isolated millstone was found during a power line survey between the McGuire and Baker quarries (Hockensmith 2000). This specimen is 92 cm in diameter and 13 to 19 cm thick (see Appendix G). The eye is 18 to 19 cm in diameter. It had edge breaks and was apparently undercut during the shaping process.

Drilled Boulders

At the McGuire Quarry, 66 boulders with drill holes were documented (Table 4). The boulders had a tremendous size range: length: 30 cm–2.90 m, width: 20 cm–2.40 m, and thickness: 15 cm–1.40 m. However, most of the boulders clustered as follows: length: 50 cm–1.66 m, width: 35 cm–1.10 m, and thickness: 25–60 cm. There are some variations in boulder sizes between the Upper and Middle quarry areas and the Lower quarry areas. In the Upper and Middle quarries, ranges were as follows: length: 30 cm–2.40 m (cluster 50 cm to 1.66 m), width: 20 cm–1.70 m (cluster 40 cm to 1.10 m), and thickness: 22–80 cm (cluster 30 to 60 cm). In the Lower quarries, ranges were as follows: length: 60 cm–2.90 m (cluster 60 cm to 1.30 m), width: 32 cm–2.40 m (cluster 35 cm to 1.00 m), and thickness: 15 cm–1.40 m (cluster 25 to 60 cm). A major distinction between the boulders is that many of the specimens from the upper area are quarried slabs while all the specimens from the lower quarry area are natural boulders.

A total of 140 drill holes were recorded (Figure 23). These holes can be subdivided by diameter as follows: 2.5 cm (n=1), 3 cm (n=26), 3.5 cm (n=94), 4 cm (n=18), and 4.5 cm (n=1). At the Upper and Middle quarries, the 3.5 cm diameter holes are the most common, followed by 3.5 cm holes and a few 4 cm holes. In the Lower Quarry areas, the 3.5 cm diameter holes are the most common; some 4 cm holes are present, while 3 cm holes are rare. With the exception of one complete hole, the holes were cross-sectioned lengthwise (Figure 24). The holes can be further subdivided as 100 vertical and 40 horizontal. Drill hole depth ranged as follows: 2.5 cm diameter, 7 cm deep; 3 cm diameter, 8–14 cm deep; 3.5 cm diameter,

Left: Figure 23. Boulder # 27 at the Lower McGuire Millstone Quarry. Note the drill holes. Photograph (facing southwest) taken on March 8, 1988. *Right:* Figure 24. Boulder # 35 at the Lower McGuire Millstone Quarry. Photograph (facing west) taken on April 20, 1990.

4–32 cm deep; 4 cm diameter, 3–25 cm deep; and 4.5 cm diameter, 10 cm deep. Two sizes of drill holes cluster in more restricted depth ranges: 3.5 cm diameter, 9–14 cm; and 4 cm diameter, 11–14 cm. Drill hole spacing ranges from 7 to 53 cm with most (83 percent) clustering between 14 and 30 cm. The distance between the end holes and the boulder edges ranges between 8 and 54 cm with most (75 percent) clustering between 15 and 32 cm. The fracture types produced by the drill holes were as follows: straight (n=28), concave (n=25), irregular (n=12), and convex (n=2).

Some of the boulders may actually be millstone preforms. These include boulders # 12 (70 × 130 × 60 cm), # 19 (240 × 100 × 45 cm), # 21 (100 × 90 × 42+ cm), # 33 (230 × 185 × 100 cm), # 34 (135 × 80 × 40 cm), # 41 (185 × 95 × 90 cm), # 43 (110 × 100 × 35 cm), # 50 (110 × 110 × 45 cm), # 54 (110 × 110 × 60 cm), # 57 (150 × 100 × 90 cm), # 58 (240 × 120 × 55 cm), # 59 (145 × 130 × 30 cm), # 63 (130 × 100 × 60 cm), # 64 (240 × 210 × 30+ cm), and # 66 (130 × 70 × 70 cm). These reassessments are based on information contained on the recording forms and field sketches.

Table 4. Drilled Boulder Measurements at the McGuire Quarry

Specimen No.	Length (cm)	Width (cm)	Thickness (cm)	No. of Holes
01	55	40	22	1
02	88	55	37	1
03	65	33	40	3
04	52	35	32	1
05	30+	25	30	1
06	60	20+	40	2
07	65	45	45	1
08	65	35+	40	1
09	50	35	25	1
10	80	65	50	2
11	75	55	55	2
12	130	70	60	1
13	70	40	40	1
14	75	60	55	1
15	120	60	60	1
16	150	140	60	14
17	125	63	22	1
18	155	100	35	1
19	240	100	45	5
20	82	26	32	3
21	100	90	42+	3
22	130	90	55	3
23	180	100	80	1
24	165	90	70	3
25	120	72	35	4
26	85	40	30	1
27	85	54	35	3
28	200	170	30	2
29	194	75	55	3
30	166	100	50	1
31	115	110	55	3
32	66	20	42	1
33	230	185	100	2
34	135	80	40	1
35	100	40	140	5
36	112	85	85	2

Specimen No.	Length (cm)	Width (cm)	Thickness (cm)	No. of Holes
37	80	60	45	1
38	105	88	45+	1
39	120	75	65	1
40	82	80	100+	3
41	185	95	90	2
42	80	40	30	2
43	110	100	35	1
44	90	75	25+	2
45	70	45	25	1
46	125	40	92+	1
47	120	67	30+	2
48	80	75	30	1
49	75	32	55	4
50	110	110	45	3
51	62	35	15+	1
52	90	40	35	2
53	70	60	20+	1
54	110	110	60	1
55	65+	50	50	1
56	65	45	25+	1
57	150	100	90	2
58	240	120	55	4
59	145	130	30	2
60	60	35+	50	1
61	95	50	55	4
62	290	240	120+	NA
63	130	100	60	2
64	240	210	30+	4
65	75	60	50+	2
66	130	70	70	3

+Bottom buried and maximum thickness unavailable.

While many of the boulders from the McGuire Quarry are very interesting, three boulders warrant individual discussion. Boulder # 16 (Figure 25) is a large quarried stone block measuring 150 cm long, 140 cm wide, and 55–60 cm thick. On the visible areas of this block, 14 drill holes were recorded on different surfaces. Vertical drill holes on the bottom of the block indicate that some type of derrick or large hoist was utilized to lift and turn the stone over. Most of the drill holes were straight into the stone, but several at the base on the south end of the block were drilled at an angle. This is the best example of a quarried block that was recorded at the quarry. The second interesting specimen is Boulder # 21 (Figure 26) which is 100 cm long, 90 cm wide, and 42+ cm thick. This rectangular slab has most of a leveling cross on the upper surface. The long axis of the cross extends across the entire surface (east-west), ranges in

Figure 25. South side of Boulder # 16 at the Upper McGuire Millstone Quarry. Note the drill holes on the base of the boulder. Photograph (facing west) taken on April 2, 1987.

width from 22 to 24 cm, and is 2 cm deep. One side arm was started on the northern edge (north-south). It is about 24 cm long, 20 cm wide, and 2 cm deep. No evidence exists for the other cross side arm. Two linear tool marks form the edges of the side arm of the cross. On the southwest quadrant of the upper surface, two additional linear tool marks were present. The third specimen of interest is Boulder # 62, a massive boulder with work only on one corner. This specimen is 2.9 m long, 2.4 m wide, and 1.2 m thick. A wedge shaped segment of this boulder was removed from the southeast corner. The area removed was 110 × 110 cm. At the north corner, the depression is 54 cm deep; it is 30 cm deep on the southern end, and 15 cm deep at the east edge. Two vertical tool marks are present on the northern edge of the depression and three additional vertical tool marks on the surface below the cut. It is not known whether a slab was removed for a millstone or whether the stone was opened to inspect its quality.

Figure 26. Boulder # 21 at the Upper McGuire Millstone Quarry. Note the unfinished leveling cross on the upper surface of this millstone preform. Photograph (facing southwest) taken on March 22, 1988.

Tool marks (n=12), in addition to drill holes, were observed on four boulders (Numbers 21, 29, 54, and 62). These were very shallow (1–2 cm) linear depressions cut into the sides and tops of boulders. They range between 13 and 25 cm in length. The examples on the sides of boulders were vertical and parallel to one another. The examples on a boulder (#21) top appeared to be associated with the early stage of a leveling cross.

Baker Quarry

The Baker Millstone Quarry (15Po304) is located on a slope of Rotten Point knob. Fieldwork was conducted on March 22, 23, 29, 30, 31 and April 5, 8, 1988. The upper quarry area extends from 1,200 feet (above MSL) and descends to the narrow floodplain below (900 feet above MSL). Spatially, the quarry extends over an irregular area approximately 255 m north-south and 550 m east-west (14 ha). Quarrying remains are located on four ridges with steeply sloping sides and within the intermittent streambeds between the ridges. These ridges were designated A, B, C, and D for recording purposes. Likewise, the small intermittent streams between these ridges were also assigned letter designations.

These east-west oriented ridges are actually projections originating from the slope of Rotten Point. Ridges A and C are the longest with shorter Ridges B and D between them. Ridge A is the longest and southernmost of the four ridges (Figure 27). An intermittent stream (Stream A) flows along its southern boundary while a more substantial stream (Stream B) flows along its northern slope. The quarrying remains are scattered over an area about 315 m east-west along the crest and slopes of the ridge. At the east end of the quarry, the ridge is about 126 m wide, but it gradually tapers to approximately 30 m in width at the west end. The quarrying remains increase in density towards the west with the greatest concentration along the westernmost 70 m. Quarrying remains on Ridge A include six shallow pits (Figure 28).

Ridge B is immediately north of Ridge A. It is much shorter, steeper, and only about

Figure 27. Boulders on Ridge A at the Baker Millstone Quarry. Photograph (facing east) taken on March 30, 1988.

Figure 28. Pit with boulder on west end of Ridge A at the Baker Millstone Quarry. Photograph (facing west) taken on April 8, 1988.

half the length of Ridge A. Stream B flows along its southern edge while Stream C flows along its northern edge. The archaeological remains extend about 145 m east-west along the ridge. The ridge is about 30 m wide near the eastern end but constricts to about 10 m at the western end. Most of the quarrying remains are concentrated on the western end of the ridge and along Stream C. Six shallow pits (Figure 29) are among the quarrying remains on Ridge B.

Ridge C is a long, narrow ridge, which nearly equals Ridge A in length. Stream B flows between these ridges and the ridges are separated by a 15 m wide floodplain at their western ends. The quarrying remains extend from the western end to approximately 165 m east where the remains abruptly end. The ridge is about 40 m wide at the eastern end of the quarry but rapidly contracts until the western end is about 15 m wide. The archaeological remains are fairly dense across the ridge crest. Nineteen pits, both oval and linear, were present on Ridge C.

Ridge D is the northernmost extent of the quarry as well as the shortest of the four ridges. It is very narrow, has steeply sloping sides and abruptly terminates with a very steep slope. Stream C flows between its southern slope and Ridge B. Stream D, a small intermittent stream, flows along the northern slope of Ridge D. The archaeological remains extend over an area 125 m east-west and about 38 m north-south. The quarrying remains along the ridge crest are sparse with most of the remains occurring in the streambeds. No pits were encountered on Ridge D.

The streams mentioned above are for the most part natural drains, which usually flow after rains. The exception is the lower segment of Stream B which appears to flow year round. The upper reaches of the streams are usually no more than one meter in width with steeply

Figure 29. Quarried boulder in pit on Ridge B at the Baker Millstone Quarry. Notebook at the base of the boulder serves as a scale. Photograph (facing east) taken on April 5, 1988.

sloping banks. These streams widen out to about two meters as they descend towards the bases of the ridges. The density of conglomerate boulders and fragments varies greatly between these streams and segments of the same stream. In some areas, the conglomerate is either absent or in the form of very small chunks. Elsewhere, large boulders completely cover the streambeds. Those segments of the streambeds containing large boulders also exhibit the greatest amount of quarrying remains. Stream A contains archaeological remains scattered along about 218 m of its course at the southern base of Ridge A. Stream B contain quarrying remains only along its western end between Ridges A and C. Worked boulders and millstones are scattered along Stream C (Figure 30) with the major concentration between Ridges B and D. Finally, Stream D contains quarrying remains in only its upper end.

Thirty-one pits were encountered at the Baker Quarry. These include six pits on Ridge A, six pits on Ridge B, and 19 pits on Ridge C. On Ridge A, the oval pits range from 2 to 6 m in diameter (2, 2, 2, 3, 6, and 6 m) and 50 cm to 1 meter in depth (50, 50, 50, 50, 100, and 100 cm). The oval pits on Ridge B vary in size from 1 meter in diameter up 3×4 m (1, 1.5, 2, 2×2, 2×3, and 3×4 m) across with depths ranging from 30 to 50 cm (30, 30, 50, 50, and 50 cm). Sixteen oval pits and three linear pits were located on Ridge C. The oval pits ranged in size from 1 m to 4×5 m (1, 1, 1, 1, 1.5, 1.5, 1.5, 1.5, 2, 2×3, 2×4, 2×4, 2×5, 2×6, 3×4, 4×5 m) and in depth from 30 cm to 1.5 m (30, 30, 30, 30, 30, 30, 30, 30, 30, 30, 50, 50, 50, 50 cm, 1 and 1.5 m). The three linear pits measured 1.5×16 m (75 cm deep), 1.5×5 m (30 cm deep), and 2×9 m (1 m deep).

A total of 28 millstones, 55 boulders with drill holes, and a possible grindstone blank were documented. The conglomerate worked at the quarry is a light gray to reddish brown

Figure 30. Stream C with conglomerate boulders at the Baker Millstone Quarry. Photograph (facing east) taken on April 5, 1988.

sandstone containing rounded quartz pebbles (mostly white with some yellow and brown). Most pebbles are less than 2 cm in diameter with occasional pebbles up to 4 cm in size. Other quarry remains consist of shallow pits and fragments of conglomerate from shaping millstones.

Millstones

Twenty-eight millstones were documented at the Baker Quarry. Table 5 provides a summary of the diameter, thickness, and eye diameter for each millstone from this quarry. The following discussion describes these millstones.

Millstone # 1 is a nearly complete specimen (Figure 31). The stone is 90 cm in diameter, 28 cm thick, and has an eye 17 cm in diameter that goes through the stone. Two shallow linear grooves (2 cm wide) were visible on the upper surface. These grooves may be drill hole scars from splitting the millstone from the parent boulder or associated with leveling the top of the stone. The grooves are in a radial pattern (farther apart at the outer edge and closer together towards the eye). They are 15 cm apart near the outer edge of the stone and 11 cm apart near the center. There may have been additional grooves obscured by dirt; the stone was very dirty even after one foot of silt was removed from the top of it. Three areas of edge damage are present. The lengths of the damaged areas measure 10 cm, 18 cm, and 32 cm. It would not have been feasible to have further reduced the entire diameter.

Millstone # 2 is an oval shaped preform (Figure 32). The specimen was rounded. The stone measures 100 × 110 cm and is 32 cm thick. The upper surface was in the process of being flattened. A T-shaped trough is present on the upper surface. The main trough (100 cm long and 5 cm deep) crosses near the center of the millstone preform. It is 13 cm wide across most of the stone but flares to 32 cm wide at one end. The short arm of the trough is ca. 37 cm long and less than 30 cm wide (width and depth not recorded). The stone was rejected due to shaping errors.

Figure 31. Photograph of Millstone # 1 at the Baker Millstone Quarry. Photograph (facing west) taken on March 23, 1988.

Figure 32. Photograph of Millstone # 2 at the Baker Millstone Quarry. Note the leveling cross groove on the upper surface. Photograph (facing southwest) taken on March 23, 1988.

Table 5. Millstone Measurements at the Baker Quarry

Specimen No.	Diameter/Size (cm)	Thickness (cm)	Eye Diameter (cm)
01	90	28	17
02	100 × 110	32	NA

Specimen No.	Diameter/Size (cm)	Thickness (cm)	Eye Diameter (cm)
03	82 × 92	45	NA
04	92 × 115	38	NA
05	83	32	NA
06	90 × 100	60	NA
07	75	20	NA
08	106	47	NA
09	82 × 122	45	NA
10	87	48	NA
11	100 × 118	35	NA
12	110 × 126	45	NA
13	110	28	NA
14	78	35	NA
15	80	30	NA
16	104	40	NA
17	91	40	NA
18	88 × 100	40	NA
19	88	27	17
20	100 × 103	25	NA
21	110 × 130	55	NA
22	95	26+	NA
23	140 × 170	44+	NA
24	85	15	NA
25	95	50	NA
26	91	32	NA
27	103	40	NA
28	95 × 110	50	NA

+Bottom buried and maximum thickness unavailable.

Millstone # 3 is a roughly shaped oval preform made from a boulder. The stone measures 82 × 92 cm and is 45 cm thick. The upper surface is relatively flat except for one edge. The damaged portion of the stone is about 45 cm long, up to 24 cm wide, and is 5–12 cm lower than the remainder of the upper surface. Two fractures are also present on two other edges of the specimen. A horizontal linear tool mark (2 cm wide and 32 cm long) is visible on the upper surface. A vertical drill hole scar (3 cm in diameter and 27 cm deep) is on the side opposite where the edge damage was present. The stone probably broke during the shaping of the edges.

Millstone # 4 is a very roughly shaped preform. The stone measures 92 × 115 cm and is 38 cm thick. The upper surface is relatively flat. The thickness of the slab is much greater on one side than the other. A visible tool mark on the upper surface of the stone was a central point. This round hole was near the center of the specimen was 2 cm in diameter and 1.5 cm deep. Undoubtedly, this hole was cut to serve as a reference point from which the planned diameter of the stone could be outlined. Other tool marks consist of some very faint linear marks across the upper surface of the stone. The stone appears to have been rejected because of shaping problems.

Millstone # 5 is a roughly rounded slab that was abandoned during the leveling process. It is about 83 cm in diameter and is 32 cm thick. One axis of a leveling trough had been started on the upper surface. This trough was 13 cm wide and 5 cm deep for about two thirds of the way across the specimen. An irregular break across the end of the stone obliterated the leveling cross and left a depression 10 cm lower than the main surface. The millstone appears to have been rejected because of irregular breaks during edge-shaping and problems with leveling the upper surface. Horizontal linear tool marks were present on the top of the millstone.

Two of these tool marks were associated with one end of the leveling cross. They were 2 cm wide and 22 to 27 cm long. Five additional linear tool marks were at the opposite end of the stone. They were all 2 cm in diameter and ranged in length from 10 to 22 cm (10, 12, 14, 18, and 22 cm). The tool marks appear to be associated with the leveling of the upper surface of the stone.

Millstone # 6 is an irregular boulder that has been roughly shaped (Figure 33). The stone measures 90 × 100 cm and is 60 cm thick. The upper surface is irregular. A shallow hole, 3 cm in diameter, is present near the center of the specimen. In addition to leveling problems, irregular breaks during shaping process rendered the boulder unsuitable for further reduction. Six horizontal linear tool marks were present on the top of the millstone. Five of these were present on one end and a single tool mark on the opposite end. All six linear tool marks are oriented along the same axis of the stone. The tool marks on the north end of the stone are about 2 cm wide and range in length from 5 to 24 cm (5, 18, 24, and 32 cm). They were spaced between 5 and 17 cm (5, 7, and 17 cm). One vertical drill hole scar (2 cm in diameter) extended from the top of the specimen to 17 cm down the side.

Figure 33. Millstone # 6 at the Baker Millstone Quarry. Note the horizontal tool marks on the upper surface. Photograph (facing south) taken on March 29, 1988.

Millstone # 7 is a roughly rounded slab in a streambed. It measures about 75 cm in diameter and is 20 cm thick. The upper surface is relatively flat. The stone appears to have been rejected because of shaping errors occurring while rounding the edges. No tool marks were visible.

Millstone # 8 has a perfectly flat upper surface and has roughly rounded sides (Figure 34). The top is slightly more rounded than the bottom. The stone measures 106 cm in diameter and is 47 cm thick. The most obvious flaw is undercutting on one side during the shaping of the edge. The only tool marks observed were two vertical drill holes on the northern edge of the millstone. These drill hole scars are 2 cm in diameter and 16 cm apart. One hole is 8 cm deep while the other is 15 cm deep.

Figure 34. Millstone # 8 at the Baker Millstone Quarry. Photograph (facing southeast) taken on March 29, 1988.

Millstone # 9 is a large oval boulder that has been rounded about halfway around. The upper surface has been flattened. A fracture on one side may have made further shaping unfruitful. It measures about 82 × 122 cm and is 45 cm thick. The rounded area ranges from 71 to 82 cm. No tool marks were observed.

Millstone # 10 is a roughly rounded boulder with one axis of a leveling trough on the upper surface. The entire boulder has been shaped but still retains a squarish look. It is about

87 cm in diameter and is 48 cm thick. The leveling trough extended from one edge of the stone to the surface break on the opposite end. The trough was 15 cm wide and about 2 cm deep. A linear tool mark (2 cm wide) forms the western bounder of the trough. A short linear tool mark is present at the center of the cross on the south end. The stone was rendered unsuitable for further reduction because of an irregular surface break (18 × 44 cm and 7 cm deep) on one end of the upper surface and undercutting of the side while shaping the edges.

Millstone # 11 is a roughly shaped oval with an irregular upper surface. The stone measures 100 × 118 cm and is 35 cm thick. Shaping errors and cracks across one end of the stone resulted in the millstone's rejection. Also, depressions on the upper surface ruined the top of the stone. No tool marks were visible.

Millstone # 12 is a roughly rounded boulder with an irregular upper surface (Figure 35). The millstone makers began leveling the upper surface but did not complete the task. It measures about 110 × 126 cm and is 45 cm thick. Six horizontal linear tool marks were present on the top of the millstone. Five of these were parallel on the south end and a single tool mark was at a 90-degree angle at the western edge of this cluster. The tool marks range in length from 12 to 33 cm (12, 19, 20, 21, and 33 cm). They were spaced between 6 and 13 cm apart (6, 12, 12, and 13 cm apart). Undoubtedly, these tool marks were associated with leveling the top of the millstone. The stone was rendered unusable by shaping errors and some large sandstone inclusion in the grinding surface.

Figure 35. Millstone # 12 at the Baker Millstone Quarry. Note the horizontal tool marks on the upper surface. Photograph (facing southwest) taken on March 30, 1988.

Millstone # 13 is completely rounded and flattened on the upper surface. One edge is missing. This specimen is 110 cm in diameter and is 28 cm thick. It appears that the millstone broke during the final shaping. Also, a large crack is present across the center of the millstone. No tool marks were observed.

Millstone # 14 has been rounded and is still attached to the parent boulder (Figure 36). The specimen is perfectly rounded about three-fourths of the way around the circumference and the remaining one-fourth is a little irregular. The millstone is 78 cm in diameter and is 35 cm thick. The stone has been rounded between 16 and 18 cm below the upper surface. The base of the boulder is 10–15 cm wider than the shaped portion. The millstone appears to have been rejected because one edge was damaged during the shaping process. No tool marks were visible.

Figure 36. Millstone # 14 at the Baker Millstone Quarry. Note the parent boulder still attached to the millstone base. Photograph (facing west) taken on March 30, 1988.

Millstone # 15 is half of a broken specimen (Figure 37). The stone apparently broke during the shaping process. The upper surface is slightly irregular with the surface varying about 5 cm vertically. The approximate diameter is 80 cm and the stone is 30 cm thick. No tool marks were observed.

Millstone # 16 is a roughly rounded specimen and the millstone maker had started leveling the top of the stone (Figure 38). This specimen has an approximate diameter of 104 cm and is 40 cm thick. One axis of a leveling trough is present on the eastern edge of the stone. The surviving section of the trough is 41 cm long and ranges in width from 12 cm at the center to 35 cm at the outer edge. The trough depth ranges from 3 to 5 cm. The northwestern quarter of the stone had been leveled to a depth of 5 cm. During the shaping of the edges, an old fracture or bedding plane caused the stone to break in an angular way. This break may have occurred when the upper surface was being leveled. Also, some other edge damage was apparent. These flaws made it unfeasible to continue reducing the millstone.

Figure 37. Photograph of Millstone # 15 at the Baker Millstone Quarry. Photograph (facing west) taken on March 30, 1988.

Figure 38. Photograph of Millstone # 16 at the Baker Millstone Quarry. Photograph (facing east) taken on March 30, 1988.

Millstone # 17 is half of a nearly complete millstone that was purposely split in half. It is 91 cm in diameter and is 40 cm thick. The sides are completely rounded and the upper surface is flat. This is the only example at the six millstone quarries of a millstone purposely split in half. Two holes were drilled from the upper surface of the millstone to split it. The holes are 33 cm apart. Both holes are 3.5 cm in diameter and 10 cm deep. Only minor edge damage (10 × 3 cm and 2 cm deep) is apparent along one edge and two large chert inclusions were observed that could have made the stone undesirable.

Millstone # 18 is a rectangular slab with rounded corners. The upper surface is relatively flat. The millstone measures 88 × 100 cm and is 40 cm thick. Five linear horizontal tool marks were present on the upper surface of the specimen. They range in length from 17 to 32 cm (17, 17, 22, 24, and 32 cm long). Three parallel tool marks are 19 and 22 cm apart. They were probably associated with leveling the upper surface of the millstone. Since only the top of this specimen was exposed, it was not possible to see the flaws that resulted in its rejection.

Millstone # 19 is a completed millstone that broke into two halves (Figure 39). It is 88 cm in diameter and is 26 cm thick. The stone is perfectly rounded and flattened on top. It has a central eye that is 17 cm in diameter that goes through the stone. The stone may have

Figure 39. Millstone #19 at the Baker Millstone Quarry. This millstone probably broke during the cutting of the eye. Photograph (facing west) taken on April 5, 1988.

split in two sections during the final stage of cutting the eye. A small portion of the upper section is missing. No tool marks were observed.

Millstone # 20 is a rectangular slab that has been roughly rounded. The upper surface varies 4 to 5 cm vertically. The specimen measures 100 × 103 cm and is 25 cm thick. Two vertical tool marks are present on the western edge of the stone. One tool mark is 2 cm wide and 18 cm long. The second tool mark is 3 cm wide and 12 cm long. Irregular breaks in shaping the sides made the stone too small to further shape.

Millstone # 21 is roughly rounded but still retains a rectangular form. This boulder measures 110 × 130 cm and is 55 cm thick. The western side of the boulder was shaped down 21 to 32 cm from the upper surface with the base of the stone being wider. A single vertical tool mark was observed. The tool mark was 32 cm long and had a width ranging between 3 and 4 cm. Additional marks may have been present but the stone was very muddy. The specimen appears to have been rejected because of irregular breaks during the shaping of the sides.

Millstone # 22 has been rounded except for a small projection (12 cm long and 26 cm wide) on the northeast edge. The upper surface of the millstone is very flat. It measures about 95 cm in diameter and is over 26 cm thick. The stone was rejected since the base was undercut (ca. 15 cm) on one side during the shaping process. No tool marks were observed.

Millstone # 23 is an oval shaped boulder with initial surface leveling. This boulder measures 140 × 170 cm and is 44+ cm thick. Four linear horizontal tool marks (all 5 cm wide) were observed on the upper surface of the specimen. They range in length from 15 to 50 cm (15, 47, 50, and 50 cm long). The three parallel tool marks are 22 and 50 cm apart. They were probably associated with leveling the upper surface of the millstone. A single horizontal drill hole was also present. It was 3 cm in diameter and 10 cm deep. This millstone was rejected because of irregular breaks during the shaping of the sides and a surface break (20 × 56 cm and 15 cm deep).

Millstone # 24 is a roughly rounded slab with a slightly irregular upper surface. It is 85 cm in diameter and 6 to 15 cm in thickness. Flaws include minor shaping errors and irregular thickness (6–15 cm). No tool marks were observed.

Millstone # 25 is a thick, roughly rounded boulder (Figure 40). It is rounded about three-fourths around the circumference. The millstone makers had started leveling the upper surface. It measures about 95 cm in

Figure 40. Millstone # 25 at the Baker Millstone Quarry. Photograph (facing east) taken on April 5, 1988.

diameter and is 50 cm thick. Five linear horizontal tool marks (2–3 cm wide and 1–2 cm deep) were present on the upper surface of the specimen. Three were oriented north-south and two were oriented east-west. Also, they were spread out so that one tool mark was in each quadrant of the circle except for the southeast, which had two tool marks. They range in length from 20 to 36 cm (20, 26, 32, 35, and 36 cm long). They were probably associated with leveling the upper surface of the millstone but could also be associated with removing a millstone from the parent boulder. Trimming errors on the edges and undercutting of the base made the stone too flawed to continue reducing.

Millstone # 26 is a completely rounded slab but the sides are not perfectly vertical. Portions of the upper surface are irregular. Three shallow depressions are on the upper surface (24 × 42 cm and 2 cm deep; 11 × 20 cm and 2 cm deep; and 10 × 11 cm and 3 cm deep). The millstone measures about 91 cm in diameter and is 32 cm thick. Three linear horizontal tool marks were present on the upper surface of the specimen. They are spaced so that one tool mark is on each third of the upper surface. They range in length from 22 to 27 cm (22, 23, and 27 cm long). Also, two vertical drill holes are located 26 cm apart on the southern edge of the specimen. Both drill holes are 3 cm in diameter and 7 cm deep. In addition to the surface irregularities, there may be other flaws that are obscured by soil covering half the base. The tool marks may be associated with leveling the upper surface of the millstone but could also be associated with removing a millstone from the parent boulder.

Millstone # 27 is a roughly rounded slab that is perfectly flat on the upper surface. Approximately one-third of the stone broke off during the shaping of the millstone. It is 103 cm in diameter and is 40 cm thick. No tool marks are visible.

Millstone # 28 is an oval slab that represents an early stage specimen. The upper surface is relatively flat. This stone measures 95 × 110 cm and is 50 cm thick. Shaping errors on the edge made the millstone unsuitable for further reduction. Tool marks were not visible on exposed portions of the stone.

Miscellaneous Specimen # 1 is not a conglomerate millstone. Since the raw material is light brown sandstone, this specimen may be a preform for a large grindstone. Conglomerate boulders were located nearby in Stream B. This rounded slab has a relatively flat upper surface. It is 80 cm in diameter and is 26 cm thick. About half of the specimen is obscured by soil and the remainder is partially under water. The only tool mark on the exposed portion of the stone was a central point. The square hole in the center of the specimen is 1 × 1 cm and about 1 cm deep. Undoubtedly, this hole was to serve as a reference point from which the planned diameter of the stone could be outlined. Larry Meadows has seen some other possible grindstone blanks near the top of Rotten Point where sandstone outcrops are present. The millstone makers may have produced grindstone as a sideline. The same methods could have been employed to produce large grindstones from the much softer sandstone.

The 28 millstones from the Baker Millstone Quarry represent several stages in the manufacturing sequence. These stages include nearly complete millstones (n=3), advanced preforms (n=2), oval or rounded preforms (n=19), and rectangular preforms (n=4). They range in diameter from 75 cm to 1.40 m but most rejected millstones cluster between 80 cm and 1.10 m. In terms of thickness, they vary from 15 cm to 60 cm with most specimens clustering between 25 cm and 50 cm. Millstones were rejected for a variety of reasons including irregular breaks (n=19), undercutting (n=2), edge damage (n=2), surface depressions (n=2), cracks (n=4), splitting apart (n=1), inclusions (n=1), and unknown (n=1). Sometimes two or three flaws occurred on the same specimen.

Other noteworthy features of the Baker millstones include eyes and leveling crosses. Two leveling crosses with a "T" or "L" configuration and two crosses with a single arm completed were encountered on the millstones. The crosses range in width from 12 to 35 cm (most 13–15 cm) and in depth from 2 to 5 cm (most 5 cm). Two millstones (Numbers 1 and 19) contained eyes that were cut all the way through the millstones. Both eyes had diameters that were 17 cm.

Manufacturing evidence consists of linear tool marks and drill holes. Thirteen millstones (Numbers 1, 3, 4, 5, 6, 8, 12, 18, 20, 21, 23, 25, and 26) had linear tool marks and four millstones (Numbers 3, 17, 23 and 26) had drill holes. The horizontal linear tool marks (n=39) ranged in number from one to seven. Four specimens had a radial pattern (Millstones 1, 12, 23, and 26). Available size ranges for the horizontal marks were 5 to 50 cm long, (most 10–35 cm), 2 to 5 cm wide, 1 to 2 cm deep, and 5 to 22 cm apart (most 5–17). Vertical tool marks (n=6) occurred as either one mark or two marks. They ranged from 8 to 32 cm long, 2 to 4 cm wide (most 2 cm) and 16 cm apart. The six drill holes were 3–3.5 cm in diameter, ranged in depth between 7 and 27 cm (most 7–10 cm), and were 26–33 cm apart.

Drilled Boulders

Fifty-five boulders with drill holes were documented at the Baker Quarry (Table 6). The boulders had the following size range: length: 44 cm–1.70 m; width: 15 cm–1.18 m; and thickness: 18–85 cm. Most of the boulders cluster within smaller size ranges: length: 48 cm–1.10 m, width: 35–90 cm, and thickness: 25–60 cm.

Ninety-six drill holes were documented on the boulders. These holes can be subdivided by diameter as follows: 2.5 cm (n=2); 3 cm (n=66); 3.5 cm (n=25); and 4 cm (n=3). With the exception of two complete holes, the holes were split through the center of their long axis. The holes were further subdivided to include orientation: 78 vertical and 18 horizontal. Drill hole depth ranged as follows: 2.5 cm diameter (10 cm); 3 cm diameter (3–22 cm); 3.5 cm diameter (5–14 cm); and 4 cm diameter (14–15 cm). Two sizes of drill holes cluster at the following depths: 3 cm diameter (9–12 cm) and 3.5 cm diameter (7–14 cm). Drill hole spacing ranges from 10 to 46 cm with most (73 percent) clustering between 18 and 31 cm. The distance between the end drill holes and the boulder edges range between 10 and 55 cm with most (82 percent) clustering between 17 and 36 cm. Fracture types produced by the drill holes were as follows: straight (n=24), concave (n=13), irregular (n=11), and convex (n=5).

Some of the boulders may actually be millstone preforms. These include boulders # 5 (80 × 80 × 40 cm), # 6 (80 × 90 × 40 cm), # 7 (93 × 126 × 55 cm), # 8 (85 × 128 × 38+ cm), # 13 (90 × 110 × 40 cm), # 15 (86 × 95 × 50+ cm), # 18 (110 × 130 × 40 cm), # 19 (118 × 130 × 50 cm), # 22 (70 × 100 × 25 cm), # 27 (90 × 93 × 48 cm), # 32 (85 × 130 × 60 cm), # 36 (70 × 110 × 50 cm), # 42 (70 × 90 × 50 cm), # 46 (100 × 170 × 48 cm), and # 47 (85 × 85 × 60 cm). These reassessments are based on information contained on recording forms and field sketches.

Table 6. Drilled Boulder Measurements at the Baker Quarry

Specimen No.	Length (cm)	Width (cm)	Thickness (cm)	No. of Holes
01	84	25	35	2
02	110	62	32	3
03	80	35	45	2
04	44	42	35	1
05	80	80	40+	6

Specimen No.	Length (cm)	Width (cm)	Thickness (cm)	No. of Holes
06	90	80	40	2
07	126	93	56	5
08	128	85	38	2
09	110	40	40	1
10	100	70	22	2
11	65	60	30	2
12	55	35	26	1
13	110	90	40	3
14	53	43	20	1
15	95	86	50+	1
16	100	55	20	2
17	140	55	60	2
18	130	110	40	1
19	130	118	50	2
20	75	35	52	1
21	85	35	20	2
22	100	70	25	1
23	95	95	60	3
24	100	47	28+	1
25	126	40	60	3
26	100	55	42	1
27	93	90	48	3
28	50	20	32	1
29	60	35	18	1
30	160	50	85	3
31	75	70	45	1
32	130	85	60	3
33	85	80	60	3
34	100	72	40	3
35	160	95	70	3
36	110	70	50	2
37	55	49	32	1
38	75	38	25	1
39	105	72	48	1
40	82	30	30	2
41	48	45	30	1
42	90	70	50	2
43	80	50	38	1
44	82	60	45	2
45	60	15	30	1
46	170	100	48	1
47	85	85	60	1
48	90	45	40	2
49	100	60	35	1
50	82	42	30	1
51	100	49	35+	1
52	97	47	35+	1
53	110	42	40+	1
54	70	35	35	2
55	156	90	25	1

+Bottom buried and maximum thickness unavailable.

Tool marks (n=13) were recorded on four boulders. Most of these are narrow (2 cm) linear depressions cut into the sides and tops of boulders. They range between 8 and 20 cm apart and are parallel. On Boulder # 5 (Figure 41), several of these appear to be associated with a leveling cross. This specimen indicates that tool marks were formed by placing shallow drill

Figure 41. Boulder # 5 at the Baker Millstone Quarry. Millstone preform with tool marks on the upper surface. Photograph (facing south) taken on March 29, 1988.

holes along a straight line and then removing the stone from between the holes. Another type of tool mark was observed on Boulder # 24. This is a 67 cm long and 3–3.5 cm wide shallow (1–2.5 cm deep) depression on the top of the boulder. This may be an abrading area for tool sharpening.

Toler Quarry

The Toler Millstone Quarry (15Po306) is located on Rotten Point knob. Fieldwork was conducted on March 26 and 27, 1987; March 24 and April 12, 1989. Quarrying remains are present in three contiguous areas extending over an area 310 m north-south and 200 m east-west (6.2 ha). The most impressive area is the Upper Quarry (Figure 42), which is located on a bench (1,360 feet above MSL). At the southern end of this bench, a large depression has been excavated into the side of the knob. This quarry depression is about 30 m north-south and 15 m east-west with a back wall 9 m high on the western side. The floor of the pit is rel-

Figure 42. Main quarry area at the Toler Millstone Quarry. Photograph (facing northwest) taken on March 27, 1987.

Figure 43. Pit # 1 at the Toler Millstone Quarry. Photograph taken on March 27, 1987.

Figure 44. Pit # 2 at the Toler Millstone Quarry. Photograph taken on March 27, 1987.

atively level and contains two millstones and some conglomerate slabs and rubble. Approximately 115 m north of this large depression is an oval pit (17 × 18 m across) (Figure 43). This pit is 1 to 1.5 m deep in the southern half. However, there is an east-west oriented channel cut across the northern end of this pit that is 2 to 2.5 m deep. About 10 m further north is an irregular pit (covering an area ca. 15 × 22 m) with a maximum depth of 2.5 m (Figure 44). This pit has three projections and a maximum depth of 2.5 m. Part of the pit has narrow deep channels in the bottom.

The large depression and pits described above represent quarrying activities, which exposed in situ deposits of conglomerate. About 27 m upslope from the two pits is a conglomerate cliff about 3 m high. Concentrations of rubble indicate that the shaping debris from millstone manufacturing were piled at the edge of the pits and frequently thrown over the steep slope. The bench between the main quarry depression and the two pits may have served as a work area. This bench is about 88 m north-south and 40 m east-west.

Immediately north of the pits on the steep slope of the knob is the Northern Quarry. Six millstones, one drilled boulder, and thousands of broken pieces of conglomerate are scattered across an area approximately 100 m north-south and about 60 m east-west. This area represents both in situ working of scattered boulders on the slope and discarded rubble from the Upper Quarry. A bench in this area of the quarry is approximately 15 m wide and 60+ m long.

The third quarry area (Lower Quarry) is located immediately east (down slope) of the Upper Quarry. Quarry remains are distributed across an area approximately 110 m east-west and about 130 m north-south. This area represents the in situ working of boulders in and near an intermittent stream bed draining the north end of Rotten Point. Remains in this area include three millstones still attached to the parent boulders, seven drilled boulders, and scattered conglomerate rubble.

During the fieldwork 15 millstones in various stages of completion and 20 boulders with drill holes were documented at the Toler Quarry. The conglomerate used for millstone manufacture is light gray to light tan sandstone containing rounded quartz pebbles (mostly white with some yellow, brown and red). Most of the pebbles are between 1 and 2 cm in diameter with rare examples up to 7 cm in diameter. Large chert inclusions are also present in the conglomerate.

Millstones

Fifteen millstones were documented at the Toler quarry. Table 7 provides a summary of the diameter, thickness, and central hole diameter for each millstone from this quarry. The following discussion describes these millstones.

Millstone # 1 is an advanced stage millstone that cracked during the manufacturing process (Figure 45). It is 108 cm in diameter and is 25 to 40 cm in thickness. The eye in the center of the stone is 16 cm in diameter but was only cut to a depth of 15 cm. The stone was rounded 10 to 15 cm below the upper surface with the parent boulder remaining somewhat irregular. Apparently, the millstone cracked before it could be separated from the parent boulder.

Figure 45. Millstone # 1 at the Toler Millstone Quarry. Photograph taken on March 26, 1987.

Millstone # 2 is an advanced stage specimen (Figure 46) that that was damaged during the manufacturing process. It is 75 cm in diameter and is 16 to 18 cm thick. The eye in the center of the stone is 14 cm in diameter and was cut through the entire stone. The stone was roughly rounded but still somewhat irregular on the upper surface. Four linear, but not straight, grooves are present on one side of the upper surface. They range between 9 and 32 cm in length and between 1.5 and 2 cm in width. The millstone appears to have been rejected because of edge damage during the trimming of the sides. There are also some cracks on one side.

Figure 46. Millstone # 2 at the Toler Millstone Quarry. Photograph taken on March 26, 1987.

Millstone # 3 is an advanced stage millstone (Figure 47) that was damaged during the manufacturing process. It is 150 cm in diameter and is 35 to 42 cm thick. The eye in the center of the stone is 24 cm in diameter but was only cut to a depth of 11 cm. The stone was shaped to 18 to 20 cm below the upper surface with the parent boulder flaring outward 7–8 cm in an irregular fashion. About 21 cm of the unshaped parent rock was visible and the remainder was buried in the soil. The millstone is well rounded and the upper surface is relatively flat. The millstone appears to have been rejected because of edge damage in several places during trimming of the sides, as well as undercutting and a crack. An area 70 cm long, 18 cm wide, and 5–15 cm thick spalled off the upper surface and a spall came off the opposite side that was about 50 cm long, 8 cm wide, and 8 cm thick. An unusual feature of this millstone was a keystone-shaped piece that was cut from the upper surface but left in place (Figure 48). At the eye, this piece was 26 cm wide and expanded to 42 cm at the other end. The piece was 40 cm long and was formed by cutting a 1 cm wide groove. It would have taken a special tool to cut such a narrow groove in the conglomerate. The piece resembles a standard piece used to construct a composite millstone. The reason for cutting this piece remains a mystery.

Figure 47. Millstone # 3 at the Toler Millstone Quarry. Photograph taken on March 26, 1987.

Figure 48. Close-up view of keystone-shaped cut on Millstone # 3 at the Toler Millstone Quarry. Photograph taken on March 26, 1987.

Millstone # 4 is nearly complete specimen (Figure 49) that was damaged during the manufacturing process. It is 110 cm in diameter and is 30 cm thick. The eye in the center of the

Figure 49. Millstone # 4 at the Toler Millstone Quarry. Photograph taken on March 26, 1987.

stone is 16 to 18 cm in diameter and was cut through the entire stone. The hole is not perfectly round and expands towards the bottom. The stone was rounded but has a rough upper surface. The millstone appears to have been rejected because of edge damage and cracking. A large piece measuring about 70 cm long, 30 cm wide, and 12–18 cm thick spalled off the upper surface of the millstone.

Table 7. Millstone Measurements at the Toler Quarry

Specimen No.	Diameter/Size (cm)	Thickness (cm)	Eye Diameter (cm)
01	108	40	16
02	75	18	14
03	150	42	24
04	110	30	18
05	87	25	16
06	92	20* (44)	17
07	85	20* (32)	17
08	90 × 96	38	18
09	160	60	NA
10	63	40	NA
11	90 × 105	45+	NA
12	60 × 70	30+	NA
13	53 × 80	30+	NA
14	50 × 57	15	NA
15	90 × 110	25	NA

*Thickness of shaped portion.
+Bottom buried and maximum thickness unavailable.

Millstone # 5 is an advanced stage millstone (Figure 50) that was damaged during the manufacturing process. It is 87 cm in diameter and is 25 cm thick. The eye in the center of the stone is 16 cm in diameter and was cut through the entire stone. The stone was rounded but has a rough upper surface. The millstone appears to have been rejected because of edge damage. A large piece measuring about 34 cm long, 11 cm wide, and 10 cm thick spalled off the upper surface of the millstone. Also, three large depressions were located on the upper surface. These depressions may represent inclusions that eroded out after the stone was abandoned or came out during the manufacturing process.

Figure 50. Millstone # 5 at the Toler Millstone Quarry. Photograph (facing south) taken on March 27, 1987.

Millstone # 6 is an advanced stage millstone (Figure 51) that broke during the manufacturing process. It is 92 cm in diameter and is 44 cm thick. The eye in the center of the stone is 17 cm in diameter but was only cut to a depth of 18 cm. The stone was rounded 18 to 20 cm below the upper surface while the parent boulder is still somewhat irregular. Possible trimming fragments are nearby. Two drill holes are on the unfinished base of the specimen. The holes are 3 cm in diameter, 18 cm long, and 18 cm apart. More drill holes may be present on obscured portions of the base. It appears that the millstone maker attempted to split the millstone from its unfinished base. However, the stone split at an angle leaving one edge very thin and removing part of one edge. Also, the millstone cracked in the same general area.

Figure 51. Millstone # 6 at the Toler Millstone Quarry. Photograph taken on March 27, 1987.

Millstone # 7 is an advanced stage specimen (Figure 52). The stone measures 85 cm in diameter and ranges in thickness between 15 and 32 cm. The eye in the center of the stone is 17 cm in diameter and goes through the millstone. The stone was rounded 16 to 20 cm below the upper surface of the parent boulder. The parent boulder has been rounded but is still somewhat irregular. Possible shaping debris is located nearby. It appears that the millstone maker attempted to split the millstone from its parent boulder. However, the stone split at an angle leaving one edge very thin (10–15 cm). Also, the millstone has a shallow depression on the upper surface.

Figure 52. Millstone # 7 at the Toler Millstone Quarry. Photograph taken on March 27, 1987.

Millstone # 8 is an advanced stage specimen (Figure 53) that was rejected late in the process. The diameter ranges between 90 and 96 cm in diameter, and the stone is 38 cm thick. The eye in the center of the stone is 17 cm in diameter but was only cut to a depth of 17–18 cm. The stone was rounded 18 to 22 cm vertically below the upper surface with about 20 cm parent boulder still remaining on the base. This stone is unique in that it was rounded slightly more than

Figure 53. Millstone # 8 at the Toler Millstone Quarry. Photograph (facing south) taken on March 27, 1987.

halfway around the circumference. The edges flare out slightly where the shaping was stopped. A vertical drill hole (2 cm in diameter and 15 cm long) was present 6 cm beyond the shaped area on the unshaped portion of the millstone. In this case, the millstone maker apparently flattened the upper surface, cut the eye, and then started rounding the stone. No obvious flaws were observed in the stone but much of the surface was covered with moss. It is also possible that the specimen's diameter was not sufficient to continue rounding.

Millstone # 9 is an oval slab that has been rounded. The approximate diameter is 160 cm and the stone is 60 cm thick. The millstone makers had started leveling the upper surface, which varies in elevation between 5 and 10 cm. Two horizontal drill holes and five linear tools marks were present. The drill holes are about 25 cm vertically below the upper surface and appear to be associated with the removal of excess stone from one side. The drill holes are 3.5 cm in diameter, 21 cm apart, and range in depth between 8 and 12 cm. The five linear tool marks on the upper surface occur in two groups, three on the west side and two on the east side. They are in a radial pattern. The western tool marks are 2–3 cm wide, 21 to 30 cm long (21, 29, and 30 cm), and 26 to 29 cm apart (26 and 29 cm). The eastern two tool marks were 15 and 25 cm long and 2 and 3 cm wide. An off centered depression on the upper surface measured 5 × 11 cm in size and was 2 cm deep. The tool marks were probably associated with leveling the upper surface of the millstone. Trimming errors on the edges and irregular breaks made the stone too flawed to continue reducing.

Millstone # 10 is a mid stage specimen (Figure 54) that is 63 cm in diameter and is 40 cm thick. The upper surface is very flat and the sides have been rounded. During the shaping of the sides, the stone broke in an irregular manner which undercut the northern edge.

Millstone # 11 is an oval slab measuring 90 × 105 cm and is 45 cm plus in thickness. The corners have been rounded and the upper surface is irregular with the elevation varying 4–5 cm vertically. Shaping errors and some undercutting on the sides rendered the stone unusable.

Millstone # 12 is an oval slab that measures 60 × 70 cm and is 30 cm plus in thickness. The stone has had some initial rounding. The upper surface is irregular with the elevation varying 5–7 cm. Shaping errors and some undercutting on the sides render the stone unusable.

Millstone # 13 represents about two-thirds of a millstone that broke during the manufacturing process. The intended diameter was about 80 cm and the stone is 30 cm thick. The upper surface is relatively flat. The millstone appears to have split during the shaping of the sides resulting in nearly one-third of the stone being broken off. Also, an edge break is present on another edge of the stone.

Millstone # 14 is a small oval slab (Figure

Figure 54. Millstone # 10 at the Toler Millstone Quarry. Photograph (facing north) taken on April 13, 1989.

Figure 55. Millstone # 14 at the Toler Millstone Quarry. Photograph (facing west) taken on March 24, 1989.

55) that measures 50 × 57 cm and is 15 cm thick. The top is flattened and the specimen is partially rounded. No flaws were observed in the stone but it appears to be too small for further reduction. This is the smallest millstone blank found at the Toler Quarry.

Millstone # 15 is an oval slab measuring 90 × 110 cm and is 25 cm thick. The slab is roughly shaped and has an uneven upper surface. The highest portion of the upper surface appears to be the original weathered surface of the stone while an area 20 cm wide along the western edge has been taken down vertically 5–8 cm. Shaping errors and some undercutting made the stone unsuitable for further reduction.

The 15 millstones from the Toler Millstone Quarry represent several stages in the manufacturing sequence. These stages include nearly complete millstones (n=1), advanced preforms (n=7), oval or rounded preforms (n=7), but no rectangular preforms (n=0). They range in diameter from 50 cm to 1.60 m but most rejected millstones cluster between 75 cm and 1.10 m. In terms of thickness, they vary from 15 to 45 cm with most specimens clustering between 25 and 45 cm. Millstones were rejected for a variety of reasons including irregular breaks (n=10), undercutting (n=3), edge damage (n=5), surface depressions (n=4), cracks (n=5), and unknown (n=2). Sometimes two or three flaws occurred on the same specimen.

Other noteworthy features of the Toler millstones include eyes. No leveling crosses were encountered on the rounded millstones. Eight millstones (Numbers 1, 2, 3, 4, 5, 6, 7, and 8) contained eyes. Four of the eyes were cut all the way through the millstones while the remaining four eyes were 11 to 18 cm deep. The diameters of the eyes ranged from 14 to 24 cm with six examples between 16 and 18 cm (16, 16, 16–18, 17, 17, and 17 cm in diameter).

Manufacturing evidence consists of linear tool marks and drill holes. Two millstones (Numbers 2 and 9) had linear tool marks and three millstones (Numbers 6, 8, and 9) had drill holes. The horizontal linear tool marks (n=9) ranged in number from four to five. Available size ranges for the horizontal marks were 9 to 32 cm long (most 21–32 cm), 1.5 to 3 cm wide, and 26 to 29 cm apart. No vertical tool marks were observed on the millstones. The five drill holes were 2–3.5 cm in diameter, ranged in depth between 8 and 18 cm, and were 18–21 cm apart.

Drilled Boulders

At the Toler Quarry, 20 boulders with drill holes were recorded (Table 8). The size of these boulders ranged as follows: length: 38 cm–1.10 m, width: 24–85 cm, and thickness: 16–70 cm. The sample was too small to identify any clustering in boulder sizes. The small size of the sample may be due in part to slabs being quarried from in situ stone rather than scattered boulders being worked. In all, 36 drill holes were documented. The holes were of three different diameters: 3 cm (n=13), 3.5 cm (n=18), and 4 cm (n=5). All the holes were split in half showing a profile along their long axis (Figure

Figure 56. Photograph of Boulder # 13 at the Toler Millstone Quarry. Photograph (facing south) taken on April 12, 1989.

56). Thirty-three of the holes were oriented vertically while only three holes were horizontal. The depth of the drill holes ranged as follows: 3 cm diameter, 7–22 cm; 3.5 cm diameter, 5–17 cm; and 4 cm diameter, 10–15 cm. Closer examination of the drill hole data indicates that the depths of most specimens cluster more tightly: 3 cm diameter, 9–12 cm; 3.5 cm, 10–15 cm; and 4 cm, 10–12 cm. Drill hole spacing in boulders ranges from 16 to 42 cm with most (77 percent) ranging between 21 and 29 cm. The distance between the end drill holes and the edge of the boulders ranges between 9 cm and 46 cm with most (74 percent) ranging between 17 cm and 30 cm. Fracture types produced by the drill holes were as follows: irregular (n=9), straight (n=6), convex (n=4), and concave (n=2).

Table 8. Drilled Boulder Measurements at the Toler Quarry

Specimen No.	Length (cm)	Width (cm)	Thickness (cm)	No. of Holes
01	96	85	60	4
02	65	34	40	2
03	60	45	30	2
04	80	27	35	2
05	42	25	24	2
06	38	35	16	1
07	100	75	30	1
08	60	58	28	3
09	78	41	36	2
10	55	45+	45	2
11	65	25+	37	2
12	40	24	35	1
13	100	50	48	2
14	110	65	30	1
15	70	50	35	2
16	110	80	42	1
17	100	50	52	2
18	82	65	25	1
19	80	35	70	2
20	73	60+	17	1

+ Bottom buried and maximum thickness unavailable.

Some of the boulders may actually be millstone preforms. These specimens include boulders # 1 (85 × 96 × 60 cm), # 8 (58 × 60 × 28 cm), # 10 (45+ × 55 × 45 cm), # 14 (85 × 110 × 30 cm), # 16 (80 × 110 × 42 cm), and # 20 (60+ × 73 × 17 cm). These reassessments are based on information from the recording forms and field sketches. No unusual or unique boulders were found at the quarry.

Two tool marks were noted on Boulder # 10. These are narrow (2 cm wide) linear shallow depressions. The tool mark on the upper surface is 8 cm long and is 19 cm from the edge of the slab. The second tool mark, on the side of the slab, is 12 cm long and starts 8 cm below the upper surface.

Ware Quarry

The Ware Millstone Quarry (15Po308) is located on the north end of Kit Point. Fieldwork was conducted on March 15, 16, 17, 22, and 23, 1989. The southern end of the quarry is approximately 920 feet above MSL and slopes northward to about 840 feet above MSL.

Spatially, the quarry extends over a general area approximately 370 m north-south and 160 m east-west (5.9 ha). Quarrying remains are located on three ridges with intermittent streams located between them. All three ridges are oriented east-west and are "stair stepped" from southeast to northwest. In other words, each ridge extends further north and west than the previous ridge. For recording purposes, the ridges were designated A, B, and C. Likewise, the small intermittent streams between the ridges were also assigned letter designations.

Ridge A is approximately 200 m north-south and ranging in width (east-west) from 62 to 93 m. Quarry activities are restricted to the southern one-third of the ridge. The remainder of the ridge lacks conglomerate boulders. Ridge B (Figure 57) is approximately 130 m east-west and about 30 m north-south. The quarry remains are scattered across the ridge but are primarily concentrated on the southwest quadrant. Ridge C measures about 96 m east-west and ranges in width from about 20 to 78 m (north-south). Remains of quarrying are concentrated on the western end of the ridge and northern half in general. Quarry remains also occur in the stream beds between Ridges A and B, and between Ridges B and C (Figure 58).

Thirty-three millstones and 30 boulders with drill holes were documented. The conglomerate at the quarry is light gray to light tan sandstone containing rounded quartz pebbles (mostly white with some yellow and brown). Most of the pebbles are under 2 cm in diameter with occasional pebbles up to 4 cm in size. Chert inclusions and occasional sandstone inclusions are present in the stone. Other quarry remains consist of many oval pits and fragments of conglomerate from shaping millstones.

Pits excavated to expose boulders were present on Ridge A and on the southern end of Ridge B. Five pits were noted on Ridge A and three pits had worked boulders in them. The

Figure 57. Ridge B at the Ware Millstone Quarry. Photograph (facing northwest) taken on March 22, 1989.

Figure 58. Stream A between Ridges A and B at the Ware Millstone Quarry. Photograph (facing southeast) taken on March 22, 1989.

pits on Ridge A had the following sizes (length, width, and depth): 2.5 × 2.5 × 0.75 m; 2.5 × 2.5 × 0.75 m; 3 × 3 × 1 m; 3.5 × 2.5 × 0.75 m; and 5 × 6 × 1.5 m. Three oval pits were located on Ridge B and one of these pits had millstones and two pits had worked boulders. These pits had the following dimensions: 1.5 × 1.5 × 0.25 m; 1.5 × 1.5 × 0.5 m; and 2 × 2 × 0.75 m. Fifteen pits were documented on Ridge C with the following dimensions: 1.5 × 1.5 × 0.3 m; 1.5 × 1.5 × 0.5 m; 1.5 × 1.5 × 0.5 m; 1.5 × 1.5 × 0.75 m; 2 × 2 × 0.3 m; 2 × 2 × 0.5 m; 2 × 2 × 0.5 m; 2.5 × 2.5 × 0.5 m; 3 × 2 × 1.5 m; 3 × 2.5 × 0.75 m; 3 × 3 × 0.75 m; 3.5 × 2.5 × 1 m; 5 × 4 × 1 m; 7 × 3 × 0.75 m; and 9 × 6 × 0.3 m. At least nine of these pits contained worked boulders in them.

Millstones

A total of 33 millstones were documented at the Ware quarry. Table 9 provides a summary of the diameter, thickness, and central hole diameter for each millstone from this quarry. The following discussion describes these millstones.

Table 9. Millstone Measurements at the Ware Quarry

Specimen No.	Diameter/Size (cm)	Thickness (cm)	Eye Diameter (cm)
01	90 × 95	35	NA
02	108 × 110	36	NA
03	90	25	NA

Specimen No.	Diameter/Size (cm)	Thickness (cm)	Eye Diameter (cm)
04	75 × 90	20–30	NA
05	115	30+	23.5
06	85 × 92	70	NA
07	75 × 93	36	NA
08	87 × 87	50+	NA
09	86 × 122	25	NA
10	87	52	NA
11	100	42	NA
12	80 × 108	30	NA
13	90 × 110	33	NA
14	96 × 105	35	NA
15	96	25+	NA
16	96	30	NA
17	122	32	NA
18	100 × 120	32	NA
19	70 × 100	30+	NA
20	105 × 115	40+	NA
21	136 × 140	50	NA
22	96 × 100	32	NA
23	96	20	NA
24	96 × 110	30	NA
25	96 × 108	25	NA
26	125 × 137	45	NA
27	80	25	20
28	85 × 90	35	NA
29	80	30	20
30	90 × 100	50	NA
31	80	25+	NA
32	96 × 110	25	NA
33	80 × 110	30	NA

+Bottom buried and maximum thickness unavailable.

Millstone # 1 is a pear-shaped slab that has been rounded on one end. It measures 90 × 95 cm and is 35 cm thick. The upper surface is roughly leveled and it has been rounded about halfway around the circumference. This slab may have been quarried from one of the large boulders in the adjacent pit. The stone was rejected due to irregular breaks occurring during the shaping of the edges and also has slight undercutting.

Millstone # 2 is an advanced-stage preform (Figure 59). The diameter ranges between 108 and 110 cm with a thickness of 36 cm. It has been roughly rounded with sides that are vertical but not smoothed. A complete leveling cross is present across the upper surface of the specimen. The north-south arm is 18 to 19 cm wide and 3 to 5 cm deep. The east-west arm (4 cm deep) is 10 to 16 cm wide near the center of the stone and 18 to 25 cm wide at the outer edges of the stone. When the millstone maker tried to remove the first pie-shaped quadrant between the cross arms, the stone broke in an irregular manner. This resulted in a

Figure 59. Millstone # 2 at the Ware Millstone Quarry. Photograph (facing northwest) taken on March 17, 1989.

depression 2–3 cm deeper than the base of the leveling cross. This would have necessitated the lowering of the entire upper surface. Also, the millstone is undercut on one side. These errors rendered the stone useless. This is an excellent example of how a leveling cross was laid out and the methods employed in removing the four quadrants.

Millstone # 3 is an early-stage preform that was abandoned (Figure 60). The diameter is 90 cm and it is 25 cm thick. The stone has been roughly rounded but is still crude. It appears that one axis of a leveling cross was cut across the upper surface of the specimen. This north-south arm (45 cm long) is clearly visible on the southern end. It is 12 to 13 cm wide and 4 cm deep. The northern end has already been leveled and only a single tool mark remains where the trough once existed. This specimen represents a different approach in leveling the upper surface. The northeast quadrant had been leveled and the millstone maker was trying to level the southwest quadrant. The northwest and southeast quadrants remain unleveled. Apparently, the stone broke in an irregular manner when the southwest quadrant was being removed. This break resulted in a depression covering about one-fourth of the upper surface and also lowered the southwest quadrant and adjacent area 6 to 11 cm below the leveling cross. This major error and some undercutting on the bottom resulted in the millstone's rejection.

Figure 60. Millstone # 3 at the Ware Millstone Quarry. Photograph (facing north) taken on March 17, 1987.

Millstone # 4 is an initial-stage preform that is rectangular in shape. It measures 75 × 90 cm and is 20–30 cm in thick. The upper surface is roughly leveled but varies 2 to 3 cm vertically in elevation. The edges were damaged and the stone is too thin on one edge. Also, the pebbles are also very sparse on one side of the specimen. These factors made the stone unsuitable for further shaping.

Millstone # 5 is an advanced stage preform (Figure 61). The upper surface had been leveled and the straightening of the sides had been started. This specimen is 115 cm in diameter and is 30+ cm thick. The millstone was shaped 16–18 cm below the upper surface. The eye was 23.5 cm in diameter and ranged in depth from 2 to 4 cm on one side and was 6 cm deep on the other side. Final edge shaping resulted in two major areas of breakage and a minor break. The largest break is 75 cm long and ranges up to 16 cm wide. It is stair stepped with vertical drops of 3, 8, and 20 cm. The second largest break is 40 cm long and 5 cm wide. There is also a hairline crack between the two major breaks. Several large chert inclusions are present in the conglomerate. These flaws made the millstone unusable.

Figure 61. Photograph of Millstone # 5 at the Ware Millstone Quarry. Photograph (facing west) taken on March 22, 1987.

Millstone # 6 is a roughly rounded specimen. It measures 85 × 92 cm and is 70 cm thick. The boulder has been rounded about two-thirds around its circumference. Initial leveling was performed on the upper surface. It appears that there was a hidden seam inside the stone that resulted in the stone's breaking in a straight line horizontally and also vertically. Apparently, the end of the boulder literally separated from the main portion. Seven tool marks are present on the upper surface. Three of the marks are parallel and may be the beginning of a leveling cross. They are all 2 cm wide and range in length from 10 to 32 cm (10, 19, and 32 cm long) and are 4 and 7 cm apart. Two marks are radial in layout. They are also 2 cm wide and are both 18 cm long. Finally, two additional tool marks are on the opposite side. They are 2 cm wide, 7 and 10 cm long, and 6 cm apart. These tool marks appear to be drill holes placed end to end in a linear configuration. Undoubtedly, they were associated with the leveling of the millstone.

Millstone # 7 is a rectangular slab with three rounded corners. The top is relatively flat and the sides are nearly vertical. It measures 75 × 93 cm and is 36 cm thick. Because of shaping errors on the edges the stone was too small to further reduce and was thus rejected.

Millstone # 8 is a rectangular slab with rounded corners. About two-thirds of the top has been subjected to initial leveling. It measures 57 × 87 cm and is 50+ cm thick. Four horizontal linear tool marks and one vertical tool mark were present. The linear tool marks are on the upper surface of the stone while the vertical tool mark is near the base of one side. Two of the marks on the upper surface are oriented north-south and two marks are oriented east-west. The spatial relationship of the tool marks suggests that they may be associated with the initial layout of a leveling cross. The northern tool mark is 2 cm wide and 20 cm long while the southern tool mark is 2 cm wide and 42 cm long. Just east of the southern tool mark is a rectangular depression that is 22 cm long, 9 cm wide, and 1 cm deep. The parallel two tool marks across the center of the specimen are both 2 cm wide and are 20 and 50 cm in length. They are 10 cm apart and are probably associated with leveling the upper surface of the millstone. The remaining vertical tool mark is 2 cm wide and 18 cm long and was probably connected with straightening the side of the stone. Shaping errors resulted in major edge damage and one-third of the top is 5 to 10 cm lower than the remainder of the stone. One large chert inclusion (7 × 24 cm) was present on the upper surface, which made the raw material less desirable.

Millstone # 9 is a roughly rounded rectangular slab. The upper surface is irregular with its elevation varying 4 to 5 cm vertically. It measures 86 × 122 cm and is 25 cm thick. Irregular fractures occurring during the shaping of the edges made the slab too small for further reduction. Also, on the eastern edge of the stone's upper surface is a depression measuring 60 cm long, 22 cm wide, and 3 to 4 cm deep.

Millstone # 10 is an advanced stage preform (Figure 62) that was damaged before it was completed. The diameter is 87 cm and it is 52 cm thick. The stone is almost perfectly rounded except for the damaged area. The irregular parent stone is attached at the north end of the stone at a depth of 35 cm below the top. A leveling cross had been laid out across the upper surface of the stone. Only the

Figure 62. Millstone # 10 at the Ware Millstone Quarry. Note the leveling cross on upper surface. Photograph (facing east) taken on March 23, 1989.

southern arm remained intact. It measures 40 cm long, 14 to 20 cm wide, and 5 to 6 cm deep. The edge of the east-west axis of the cross also remains but the adjacent areas have been leveled. Portions of the two northern quadrants have been leveled while the two southern quadrants were not leveled. It appears that that the millstone maker was removing the two northern quadrants to the elevation of the leveling cross when the stone broke in an irregular manner along the eastern edge of the stone. This break extended across an area 83 cm long, up to 24 cm wide, and 2 to 5 cm deep. Near the center of the break, a smaller depression is 5 to 20 cm deep. This major leveling error rendered the millstone unusable.

Millstone # 11 is a crudely shaped oval slab (Figure 63). It has a maximum diameter of 100 cm and is 42 cm thick. The upper surface varies 5 to 6 cm vertically. An initial stage leveling cross has been laid out on the upper surface of the slab. Eight horizontal linear tools marks appear to be associated with the leveling of the millstone. The surviving portion of the leveling cross is T-shaped (across the north end of the slab). The east-west arm (15 cm wide) of the cross was formed by three parallel linear tool marks. The tool marks range in length from 34 to 80 cm (34, 40, and 80 cm long), are 2 cm wide, 2 to 3 cm deep, and 7 to 8 cm apart. These tool marks were formed by

Figure 63. Millstone # 11 at the Ware Millstone Quarry. Note the tool marks associated with the laying out of a leveling cross on the upper surface. Photograph (facing west) taken on March 23, 1989.

drilling shallow holes in a straight line. The north arm of the leveling cross is 28 to 38 cm long, 15 to 18 cm wide, and 3 to 6 cm deep. Down the center of the north arm are the bases of shallow holes in a linear arrangement. The southeast quadrant of the stone has two parallel linear tool marks (oriented east-west) that are 2 cm wide, 10–12 cm long, and 9 cm apart. The three remaining linear tool marks are located on the western edge of the stone. These tool marks are 2 cm wide, 14–42 cm long (14, 18, and 42 cm), and 10 to 15 cm apart. Shaping errors producing edge damage made the stone unusable.

Millstone # 12 is a pear-shaped slab that has been rounded about two-thirds of the way around the circumference. The upper surface varies vertically 4 to 5 cm. It measures 80 × 108 cm and is 30 cm thick. This specimen is slightly undercut on one side, has two large holes from inclusions, a raised inclusion (6 to 7 cm high), and a large sandstone inclusion. The presence of the holes and inclusions on the upper surface made the stone undesirable. No tool marks were observed on the millstone.

Millstone # 13 is a roughly rectangular slab with rounded corners. The upper surface is very irregular with its elevation varying 5 to 15 cm. It measures 90 × 110 cm and is 33 cm thick. The upper surface has three large depressions (20 × 54 × 5–10 cm; 16 × 20 × 16 cm; and 11 × 20 × 5 cm). In addition to the irregular breaks, the conglomerate was very poorly cemented and was beginning to decompose. Thus, the stone was unsuitable for millstone manufacture.

Millstone # 14 is a roughly rounded slab (Figure 64) with evidence of initial leveling on the upper surface. It measures 96 × 105 cm and is 35 cm thick. An L-shaped portion of the leveling cross still remains on the upper surface. The northern arm is 43 cm long, 18 to 20 cm wide, and 5 cm deep. The eastern arm is 30 cm long, 20 cm wide, and 5 cm deep. It

Figure 64. Millstone #14 at the Ware Millstone Quarry. Note the leveling cross on upper surface. Photograph (facing southeast) taken on March 23, 1989.

appears that the millstone maker started leveling the southwest quadrant of the millstone. Leveling progressed well until the stone fractured in an irregular manner on the southern edge of the millstone. This irregular fracture lowered the edge of the millstone from 1 to 6 cm below the desired level. The northeast, northwest, and southwestern quadrants were not leveled. The stone also includes irregular breaks from shaping the sides and undercutting in one area (10 cm). These flaws made the stone unusable.

Millstone # 15 is a roughly rounded slab with an irregular upper surface (varies 5 to 7 cm vertically). This specimen is about 90 cm in diameter and is 25+ cm thick. Two primary factors seem responsible for the stone's rejection. First, the stone broke in an irregular fashion when the edges were shaped. These breaks made two areas on the upper surface too low: one measuring 40 × 30 cm that was 4 to 5 cm deep and a second area measuring 10 × 25 cm that 5 to 10 cm deep. Also, about half of the millstone blank has sparse quartz pebbles with some areas being almost pure sandstone. The stone was rejected because of shaping error and the poor quality of the conglomerate.

Millstone # 16 is a roughly rounded slab with initial leveling (Figure 65). It has a maximum diameter of 96 cm and is 30 cm thick. All told, twelve linear tool marks are located on the upper surface, and there are two vertical drill holes. The linear tool marks were probably associated with the leveling of the millstone. Nine tool marks are oriented north-south, one is oriented east-west, and two are oriented northeast-southwest. Most of the marks occur in pairs with spacing ranging from 6 to 16 cm (6, 8, 8, 10, 12, 12, and 14 cm apart). These tool marks are 1 cm wide and vary in length from 12 to 24 cm (12, 18, 20, 20,

Figure 65. Millstone # 16 at the Ware Millstone Quarry with Millstone # 15 in the background. Photograph (facing east) taken on March 23, 1989.

20, 20, 22, and 24 cm in long). The most obvious drill hole was 3.5 cm in diameter and 6 cm deep. A possible drill hole or chisel mark 2 cm wide was present on the edge of the specimen. On the northeast portion of the upper surface (in an area measuring 12 × 20 cm), between the two parallel tool marks (oriented northeast-southwest), the surface was lowered. Shaping errors, including undercutting, made the stone unsuitable for further reduction. Also, a large chert inclusion (15 × 24 × 10 cm) was present on the upper surface.

Millstone # 17 is a roughly rounded slab with initial leveling. It has a maximum diameter of 122 cm and is 32 cm thick. Seven linear tool marks are located on the upper surface. These linear tool marks were probably associated with the leveling of the millstone. Three tool marks are oriented north-south and four are oriented northwest-southeast. The southern three tool marks are parallel, 14 to 15 cm apart, 2 cm wide, and 12 to 26 cm long (12, 20, and 26 cm long). The tool marks on the western edge are parallel lines, 20 cm apart, 1 to 2 cm wide, and 14 to 18 cm long. The final pair of tool marks are near the center of the specimen, 12 cm apart, 1 to 2 cm wide, and 35 to 45 cm in length. The millstone was damaged by shaping errors, including slight undercutting. A major break on the northern edge of the upper surface was 46 cm long, 25 cm wide, and 5 to 10 cm deep. These flaws rendered the stone unsuitable for further reduction.

Millstone # 18 is an oval slab that is largely covered with the natural cortex. It measures 100 × 120 cm and is 32 cm thick. Only a couple of areas on the surface have been worked. The upper surface varies 5 to 7 cm vertically. A single linear tool mark is present. The horizontal mark is 30 cm long, 2 to 3 cm wide, and 2 cm deep. One side is straightened. The stone was rejected due to shaping errors.

Millstone # 19 is a rectangular slab that is rounded on two corners. The top is relatively level but varies 3 to 5 cm vertically across the entire surface and some cortex still remains. The stone measures 70 × 100 cm and is 30+ cm thick. The stone was rejected because of shaping errors on one side and an irregular break that removed one edge of the upper surface (18 to 35 cm wide and 5 to 20 cm lower).

Millstone # 20 is an oval slab with an irregular top that varies 5 to 10 cm vertically. It measures 105 × 115 cm and is 40+ cm thick. An irregular trough was cut across the upper face of the stone. The trough is 8 to 14 cm wide and 5 cm deep and is not perfectly straight. The stone was rejected due to shaping errors. The western edge of the specimen has a high area up to 32 cm wide and 5 to 8 cm high. The eastern edge has a surface break 70 cm long, up to 32 cm wide, and 5 to 10 cm lower than the surrounding surfaces.

Millstone # 21 is a roughly rounded oval slab. The upper surface is irregular and varies 5 to 15 cm vertically. It measures 136 × 140 cm and is 50 cm thick. The stone appears to have been rejected because of undercutting on one side and another area (30 × 50 cm) containing sparse quartz pebbles. The eastern half of the upper surface is 10 to 15 cm higher than the surrounding surfaces. The western edge of the specimen has an upper surface break that is 24 × 32 cm and 5 cm deep.

Millstone # 22 is a roughly rounded slab that is 96 × 100 cm and is 32 cm thick. The irregular upper surface varies 5 to 10 cm vertically. The upper surface contains four linear tool marks and one horizontal drill hole scar. Two parallel marks are near the west center of the stone. They are 12 cm apart; one measures 4 cm long and 1 cm wide while the other mark is 30 cm long and 1 to 2 cm wide. Near the northeast corner of the slab are two parallel tool marks and a drill hole. The tool marks are 7 cm apart, 10 to 15 cm long, and 2 cm wide. The drill hole scar is 3 cm in diameter and 12 cm deep. An irregular surface break 23 × 65 cm lowered the northwest quadrant by 5 to 10 cm, making the stone unusable. Also, a portion of the northern quadrant of the millstone is mostly sandstone.

Millstone # 23 is an advanced stage blank (Figure 66) still attached to the parent rock. The upper portion of the slab has been rounded from 10 to 20 cm below the top.

Figure 66. Millstone # 23 at the Ware Millstone Quarry. Note the parent boulder at the base of the millstone. Photograph (facing south) taken on March 23, 1989.

The parent rock is still present on about half of the millstone base. The initial leveling of the upper surface was started but there is still a 4 to 5 cm difference in elevation. This specimen is 96 cm in diameter and is 20 cm thick. There are 15 linear horizontal tool marks on the upper surface of the millstone and four tool marks on the parent rock below. Some of the marks are parallel while other marks are more radial and are closer together toward the center of the stone. The tool marks on the upper surface are 1 to 2 cm in width and range in length from 12 to 36 cm (12, 12, 14, 14, 14, 15, 15, 18, 20, 22, 24, 24, 32, 34, and 36 cm long). In terms of spacing, the measured sample ranged between 6 and 26 cm apart (6, 8, 9, 10, 10, 10, 12, 23, and 26 cm apart). The observed flaws included the imperfect rounding of the millstone and three cracks on the upper surface.

Millstone # 24 is a roughly rounded oval slab. It measures 96 × 110 cm and is 30 cm thick. Leveling of the upper surface was started but the surface still varies 3 to 4 cm vertically. Four linear tool marks are present on the upper surface of the stone. A single tool mark 2 cm wide and 25 cm long is in the northwest quadrant of the stone. Three tool marks are in the northeast quadrant of the stone. They are 2 cm wide and range in length from 20 to 35 cm (20, 22, and 35 cm long). The two tool marks that are parallel are 12 cm apart. Undoubtedly, these tool marks are associated with the initial leveling of the stone. The stone appears to have been rejected due to shaping errors.

Millstone # 25 has been roughly rounded but remains a slightly oval slab. The upper surface varies 4 to 5 cm vertically and still retains some of the original cortex. It measures 125 × 137 cm and is 45 cm thick. No tool marks were observed. Shaping errors appear to be the reason for the rejection of the specimen.

Millstone # 26 is a roughly rounded slab that is still slightly oval. The upper surface varies 4 to 5 cm vertically and still retains some of the original cortex. It measures 96 × 108 cm and is 25 cm thick. Six linear tool marks are present on the upper surface. Two of the marks are near the center and five marks are along one side. Five of the tool marks appear to be associated with a single arm of a leveling cross. An area 17 to 20 cm wide had been lowered 5 cm. Those marks measured were 2 cm wide and 18 to 30 cm long (18, 18, 18, and 30 cm long). The parallel marks on the edge were between 8 and 12 cm apart (8, 9, and 12 cm apart). The stone appears to have been rejected because one side of the slab was undercut during the shaping of the sides.

Millstone # 27 is an incomplete advanced stage specimen represented by two fragments. It was about 80 cm in diameter and is 25 cm thick. Approximately one-half of this millstone remained. No fragments of the missing half were observed. An eye about 20 cm in diameter was cut to a depth ranging between 12 and 14 cm. The top of the stone was very level. The stone may have broken apart when the eye was cut.

Millstone # 28 is a roughly rounded slab (Figure 67). The upper surface varies 5 to 6 cm vertically. It measures 85 × 90 cm and is 35 cm thick. Five linear tool marks are present on the upper surface and are oriented at several different

Figure 67. Millstone # 28 at the Ware Millstone Quarry. Photograph (facing east) taken on March 23, 1989.

angles. The marks are 2 cm wide and 12 to 30 cm long (12, 17, 22, 22, and 30 cm long). One mark is 2 cm deep. The western edge of the specimen has been leveled. These marks are associated with the leveling of the top of the stone. The stone was rejected because of major undercutting on one side and three areas of significant surface damage (18 × 30 cm × 7 cm deep; 15 × 25 cm × 10 cm deep; and 11 × 12 cm × 5 cm deep).

Millstone # 29 consists of about one-half of a complete specimen (Figure 68). It is 80 cm in diameter and is 30 cm thick. An eye about 20 cm in diameter was cut to a depth of 12 cm. The top of the stone is very level. This might be the missing portion of millstone # 27. There is some edge damage. The stone may have broken apart when the eye was cut.

Figure 68. Millstone # 29 at the Ware Millstone Quarry. The millstone apparently broke in half during the cutting of the eye. Photograph (facing northwest) taken on March 23, 1989.

Millstone # 30 is a rectangular slab with rounded corners. The upper surface is very irregular. The slab measures 90 × 100 cm and is 50 cm thick. It appears that the stone broke in an irregular manner while being shaped which left too little of stone to continue work. Some of the stone's edges are several cm lower than the center of the stone.

Millstone # 31 is a roughly rounded slab with an irregular upper surface (varies 5 to 10 cm vertically). It is about 80 cm in diameter and is 25+ cm thick. During the shaping of the sides, two areas of the upper surface spalled off. Along the western edge, an area 44 cm long was damaged up to 27 cm wide and 10 to 20 cm deep. A smaller area on the north edge, 12 × 15 cm was damaged from 5 to 7 cm deep.

Millstone # 32 is a rounded slab with a fairly level upper surface. This specimen was 96 × 110 cm and is 25 cm thick. The northeast edge of the stone was damaged during the shaping of the sides. An area 45 cm long and up to 13 cm wide spalled off to a depth ranging from 5 to 10 cm. No tool marks were observed.

Millstone # 33 is a roughly rectangular slab with rounded corners. The upper surface is level. The slab measures 80 × 110 cm and is 30 cm thick. Edge shaping errors made the stone too small to continue rounding.

The 33 millstones from the Ware Millstone Quarry represent several stages in the manufacturing sequence. These stages include advanced preforms (n=6), oval or rounded preforms (n=20), and rectangular preforms (n=7) but no nearly complete millstones (n=0). They range in diameter from 70 cm to 1.36 m but most rejected millstones cluster between 75 cm and 1.15 m. In terms of thickness, they vary from 20 to 70 cm with most specimens clustering between 25 and 50 cm in thickness. Millstones were rejected for a variety of reasons including irregular breaks (n=30), undercutting (n=10), edge damage (n=3), surface depressions (n=1), cracks (n=2), splitting apart (n=2), and inclusions (n=3). Sometimes two or three flaws occurred on the same specimen.

Other noteworthy features of the Ware millstones include leveling crosses and eyes. Six leveling crosses were encountered on millstones at the quarry. These include one complete leveling cross, three "T" or "L" shaped crosses, and two crosses with one arm completed. The crosses ranged in width from 8 to 20 cm (most were 12–20 cm) and were 3 to 6 cm deep (most were 5 cm). Three millstones (Numbers 5, 27, and 29) contained eyes that were partially cut. These eyes were 2 to 14 cm deep (2–4, 12, 12–14 cm deep). The diameters of the eyes ranged from 20 to 23.5 cm with two examples being 20 cm in diameter.

Manufacturing evidence consists of linear tool marks and drill holes. Twelve millstones (Numbers 3, 6, 8, 11, 16, 17, 18, 22, 23, 24, 26, and 28) had linear tool marks and two millstones (Numbers 16 and 22) had drill holes. The horizontal linear tool marks (n=74) ranged in number from one to fifteen. Two specimens had radial patterns (Millstones 17 and 23). Available size ranges for the horizontal marks were 4 to 80 cm in length (most 10–42 cm), 1 to 3 cm wide (most 2 cm), 2 to 3 cm deep, and 4–26 cm apart (most 16–17 cm). Vertical tool marks (n=6) occurred in numbers ranging from one to four. They were 18 cm in long, 2 cm wide, and between 12 and 14 cm apart. The two drill holes were 3–3.5 cm in diameter and ranged in depth between 6 and 12 cm.

Drilled Boulders

Thirty boulders were documented at the Ware Quarry (Table 10). The boulders varied tremendously in size: length: 65 cm to 2.85 m, width: 20 cm to 2.2 m, and thickness: 25 cm to 1.75 m. However, most of the boulders fell within the following ranges: length: 70 cm–1.45 m, width: 35 cm–1.10 m, and thickness: 32–80 cm. Several large boulders (Boulders 1–5) were located in a cluster (Figure 69). Of particular interest was Boulder # 1 which measured 2.85 m long, 2.2 m wide, and 1.55 m thick. This boulder contained 19 drill holes (Figure 70)

Figure 69. Boulders # 1–5 at the Ware Millstone Quarry. Boulder # 1 is to the right and Boulder # 2 in the center. Photograph (facing east) taken on March 16, 1989.

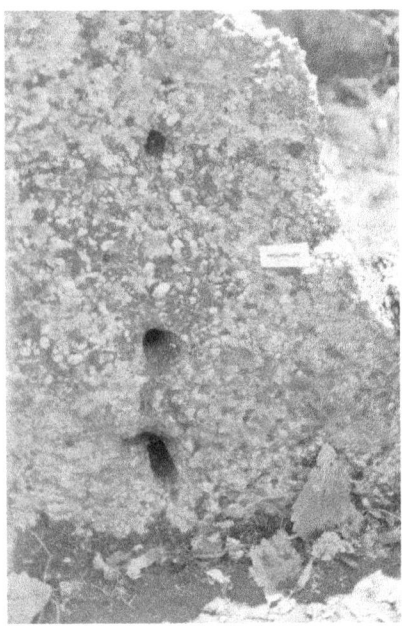

Left: Figure 70. Complete drill holes in Boulder # 1 at the Ware Millstone Quarry. Photograph (facing south) taken on March 16, 1989. *Above:* Figure 71. Tool marks on west face of Boulder # 1 at the Ware Millstone Quarry. Photograph (facing south) taken on March 16, 1989.

and 14 tool marks on two of the faces. It was the first and only specimen found that illustrated how a series (three in this case) of layers or slabs were removed from one large boulder (Figure 71).

Ninety drill holes were documented. The 88 measurable holes (Figures 72 and 73) had the following diameters: 3 cm (n=22), 3.5 cm (n=38), 4 cm (n=27), and 5 cm (n=1). A single chiseled hole had an opening size of 4 × 9 cm. Ninety-three percent (n=84) of the holes were cross-sectioned lengthwise while 7 percent (n=6) were complete holes. The holes can be further subdivided as 70 vertical and 20 horizontal. Drill hole depth had the following range: 3 cm diameter, 3–16 cm; 3.5 cm diameter, 7–15 cm; 4 cm diameter, 4–14 cm; and 5 cm diameter, 15 cm. However, drill hole depth clusters as follows: 3 cm diameter, 8–13 cm; 3.5 cm diameter, 9–12 cm; and 4 cm diameter, 10–14 cm. The spacing of drill holes ranged from 4 to 47 cm with most (68 percent) clustering between 20 and 31 cm. A related measurement, the distance between the end drill hole and boulder edge, was as follows: range 8–78 cm and

Left: Figure 72. North face of Boulder # 3 at the Ware Millstone Quarry. Note vertical drill holes. Photograph (facing southeast) taken on March 16, 1989. *Right:* Figure 73. Boulder # 10 at the Ware Millstone Quarry. Note the closely placed drill holes used to split the original boulder apart. Photograph (facing east) taken on March 17, 1989.

cluster (65 percent) 14–32 cm. The types of fractures produced by the drill holes were as follows: straight (n=9), irregular (n=8), concave (n=6), and convex (n=5).

Some of the boulders may actually be millstone preforms. These include boulders # 2 (110 × 140 × 175+ cm), # 3 (75+ × 133 × 45+ cm), # 4 (125 × 140 × 45 cm), # 5 (110 × 125 × 45+ cm), # 11 (90 × 170 × 55 cm), # 15 (200 × 155 × 80 cm), # 22 (100 × 110 × 70 cm), # 24 (85 × 107 × 25+ cm), # 28 (85 × 130 × 60+ cm), # 29 (75 × 100 × 55 cm), and # 30 (110 × 130 × 32 cm). These reassessments are based on information contained on the recording forms and in the field sketches. Boulder # 28 (Figure 74) has one axis of a leveling cross (85 cm long). The trough ranges in width 17 to 32 cm and in depth from 7 to 8 cm.

Figure 74. Boulder # 28 at the Ware Millstone Quarry. Note the presence of one axis of a leveling cross on the upper surface. Photograph (facing west) taken on March 23, 1989.

Tool marks were documented on seven boulders (Numbers 1, 2, 5, 13, 15, 18, and 29). Of these marks, 29 were vertical and 10 were horizontal. The vertical tool marks are shallow linear depressions (2 cm wide, except for one 4 cm wide tool mark). They range in length from 4 to 35 cm (most 10–20 cm long) and are spaced between 4 and 74 cm apart (most 4–16 cm). The horizontal tool marks are shallow linear depressions that are 2 cm wide. They range in length from 16 to 45 cm (most 18–30 cm long) and are spaced between 9 and 22 cm. Many of these tool marks are parallel in orientation while others are random in placement.

Table 10. Drilled Boulder Measurements at the Ware Quarry

Specimen No.	Length (cm)	Width (cm)	Thickness (cm)	No. of Holes
01	285	220	155+	19
02	140	110	175+	7
03	133	75+	45+	4
04	140	125	24–45	5
05	125	110	45+	2
06	85	40	65	1
07	65	30	35	1
08	90	75	40–55	2
09	90	50	35	2
10	110	35	45	5
11	170	90	55	3
12	98	44	40	2
13	145	45	80	2
14	88	65	70	3
15	200	155	80	2
16	80	40	40	2
17	85	35	38	1
18	110	38	52	4
19	90	20	50	2
20	90	35	50	2

Specimen No.	Length (cm)	Width (cm)	Thickness (cm)	No. of Holes
21	160	75	47+	3
22	110	100	70	1
23	70	57	32	2
24	107	85	25+	2
25	70	32	40	3
26	130	55	50	1
27	115	47+	38+	2
28	130	85	60+	NA
29	100	75	55	NA
30	130	110	32	4

+Bottom buried and maximum thickness unavailable.

Ewen Quarry

The Ewen Millstone Quarry (15Po309) is located on Kit Point between 780 and 820 feet above MSL. Fieldwork was conducted on April 24 and 25, 1990. The quarry extends over a general area approximately 150 m north-south and 70 m east-west (1.05 ha). Quarrying remains are located on the southern ends of two ridges and within the intermittent streams flanking the ridges. Both ridges are oriented north-south. While the ridges are parallel, Ridge B extends further south than Ridge A. For recording purposes, the ridges were designated A and B. The intermittent streams between the ridges were also given letter designations.

Ridge A contains the greatest concentration of quarrying remains. While the ridge is over 200 m long, only the southern 70 m of the ridge contains quarrying remains. The ridge is about 35 m wide. Stream A flows along the western base of the ridge while Stream B flows at the eastern base. Ridge A terminates in a gentle slope flanked by a small floodplain area.

Ridge B contains sparse amounts of conglomerate and thus little evidence of quarrying. Like Ridge A, Ridge B is over 200 m long but only the southernmost 50 m contain quarrying remains. The ridge is about 32 m wide. Stream A separates the eastern base of Ridge B from the western base of Ridge A. The western base of Ridge B is flanked by Stream C.

Stream A contains a number of quarry remains. Millstones and boulders were documented along the northern segment of Stream A between Ridges A and B (Figure 75) and along the segment flowing at the southeastern edge of Ridge B. Additional quarrying remains were documented downstream (south) of Ridge B in Stream A (Figure 76).

Pits excavated to expose boulders were present on Ridge A and on the southern end of Ridge B. Six pits were noted on Ridge A and two of these pits contained worked boulders in them. The pits on Ridge A had the following sizes (length, width, and depth): 2 × 2 × 0.75 m; 3 × 3 × 0.5 m; 3 × 3 × 0.75 m; 3 × 3 × 1 m; 4 × 3 × 0.75 m; and 4 × 4 × 1 m. Four oval pits were located on Ridge B and one of these pits still had a worked boulder (Figure 77). These pits had the following dimensions: 1.5 × 1.5 × 0.3 m; 2 × 2 × 0.5 m; 3 × 3 × 1 m; and 5 × 3 1.5 m.

A total of 5 millstones and 40 boulders with drill holes were documented. The conglomerate at the quarry is light gray to light tan sandstone containing rounded quartz pebbles (mostly white with some yellow and brown). Most of the pebbles are under 2 cm in diameter with occasional pebbles up to 3 cm in size. Some of the boulders are partly conglomerate and partly sandstone. Other quarry remains consist of a few oval pits and fragments of conglomerate from shaping millstones.

Figure 75. Segment of Stream A containing Millstone # 4 and several boulders at the Ewen Millstone Quarry. Photograph (facing west) taken on April 25, 1990.

Figure 76. Segment of Stream A with Millstone # 1 visible in the middle of the photograph at the Ewen Millstone Quarry. Photograph (facing southwest) taken on April 25, 1990.

Figure 77. Pit on Ridge B at the Ewen Millstone Quarry. Boulder # 40 is in this cluster of boulders. Photograph (facing northwest) taken on April 25, 1990.

Table 11. Millstone Measurements at the Ewen Quarry

Specimen No.	Diameter/Size (cm)	Thickness (cm)	Eye Diameter (cm)
01	117	40	20
02	93 × 112	28	17
03	110	40	NA
04	100	45	NA
05	88 × 90	50	NA

Millstones

Five millstones were documented at the Ewen quarry. Table 11 provides a summary of the diameter, thickness, and central hole diameter for each millstone from this quarry. The following discussion describes these millstones.

Millstone # 1 is a complete millstone (Figure 78). It is 117 cm in diameter and is 40 cm thick. The eye in the center of the stone is 20 cm in diameter and was cut through the entire slab. The upper surface of the stone was leveled and the only visible flaws were minor areas of edge damage. Since portions of the specimen are obscured by soil, other more serious defects may be hidden from view.

Figure 78. Millstone # 1 at the Ewen Millstone Quarry. Photograph (facing west) taken on April 25, 1990.

Millstone # 2 is an advanced stage preform (Figure 79). The top had been leveled but the sides are not well rounded or vertical. This oval stone measures 93 × 112 cm and is 28 cm thick. An eye 17 cm in diameter was started and cut to a depth of 8 cm. During the cutting of the eye, a large area (25 × 75 cm) extending from the eye to the outer edge of the millstone spalled off to a depth of 3 to 8 cm. There also appear to be some shaping errors on the sides.

Millstone # 3 is a roughly rounded specimen. The upper surface is irregular with vertical differences of 1 to 10 cm. The sides are also irregular. The stone measures approximately 110 cm in diameter and is 40 cm thick. The angular northwest portion of this specimen suggests that this millstone was reduced from a

Figure 79. Millstone # 2 at the Ewen Millstone Quarry. Photograph (facing north) taken on April 25, 1990.

rectangular slab. Four vertical drill holes associated with the original slab are present, two on each side of the corners. The drill hole scars are 3.5 cm in diameter and 9 to 12 cm (9, 10, 10, and 12 cm) deep. Holes # 1 and # 2 are 32 cm apart while holes # 3 and # 4 are 22 cm apart. The beginning of a possible leveling trough arm was started on the northeast potion of the stone. An area 17 cm long and 15 cm wide was leveled to a depth of 7 cm below the surrounding area. A linear tool mark 21 cm long, 1 cm wide, and 1 cm deep is present in the southern portion of the upper surface. Five vertical tool marks are present on the east side of the stone. The stone appears to have been rejected because of shaping errors, large sandstone inclusions, and a portion of the base that is nearly pure sandstone.

Millstone # 4 is roughly rounded (Figure 80) with an irregular upper surface (varies 2 to 12 cm). The sides of the stone are roughly vertical. This specimen is about 100 cm in diameter and is 45 cm thick. One horizontal drill hole scar is present on the upper surface that is 3 cm in diameter and 13 cm deep. Two linear tool marks were also on the upper surface. They were 13 to 20 cm long, 2 cm wide, and 1 cm deep. The drill hole and two tool marks are spaced so that each mark is about one-third around the circumference of the millstone. It is possible that a millstone was split off this piece. A depression 16 × 32 cm is present on one upper edge that is 15 cm deep. Only the upper 15 cm of the stone is conglomerate and

Figure 80. Millstone # 4 at the Ewen Millstone Quarry. Photograph (facing north) taken on April 25, 1990.

the remainder is pure sandstone. This specimen was rejected because of spacing errors.

Millstone # 5 is a rectangular slab with rounded corners. The irregular upper surface varies 4 to 5 cm in elevation. It measures 88 × 90 cm and is 50 cm thick. Two horizontal drill hole scars are on the west side of the upper surface. The drill holes were 3 cm in diameter, 10 to 11 cm deep, and 32 cm apart. The stone was rejected due to shaping errors and some undercutting of the base. It is also possible that a millstone was split from this slab and millstone # 5 is the discarded portion.

The five millstones from the Ewen Millstone Quarry represent several stages in the manufacturing sequence. These stages include nearly complete millstones (n=1), advanced preforms (n=1), oval or rounded preforms (n=2), and rectangular preforms (n=1). They range in diameter from 88 cm to 1.17 m with no apparent cluster of size. In terms of thickness, they vary from 28 cm to 50 cm with most specimens clustering between 40 cm and 45 cm. Millstones were rejected for a variety of reasons including irregular breaks (n=4), undercutting (n=1), edge damage (n=1), surface depressions (n=2), and inclusions (n=1). Sometimes two or three flaws occurred on the same specimen.

Other noteworthy features of the Ewen millstones include eyes. No leveling crosses were encountered on millstones at the quarry. Two millstones (Numbers 1 and 2) contained eyes. One of the eyes was cut all the way through the millstone while the other eye was 8 cm deep. The diameters of the eyes ranged from 17 cm to 20 cm.

Manufacturing evidence consists of linear tool marks and drill holes. Two millstones (Numbers 3 and 4) had linear tool marks and three millstones (Numbers 3, 4, and 5) had drill holes. The horizontal linear tool marks (n=3) ranged in number from one to two. Available size ranges for the horizontal marks were 13 to 21 cm long, 1 to 2 cm wide, and 1 cm deep. Five vertical tool marks were observed on one millstone but no measurements were recorded. The seven drill holes were 3 to 3.5 cm in diameter, ranged in depth between 9 and 13 cm, and were 22–32 cm apart.

Drilled Boulders

At the Ewen Quarry, 40 boulders were documented (Table 12). These boulders had the following size range: length: 45 cm–1.80 m, width: 27 cm–1.60 m, and thickness: 12–90 cm. While there was considerable range in boulder size, the majority fell within the following range: length: 70 cm–1.35 m, width: 35 cm–1.20 m, and thickness: 25–60 cm. These boulders contained 64 vertical drill holes and 13 horizontal drill holes. Fifty-six percent (n=43) of the holes were 3.5 cm in diameter while the remaining 44 percent (n=34) were 3 cm in diameter. All drill holes were exposed in cross-section where the stone had been split (Figure 81). Depth for the 3.5 cm holes ranged from 8 to 15 cm with most (93 percent) clustering between 10 and 14 cm. The 3 cm holes ranged from 7 to 15 cm in depth with most (85 percent) clustering between 9 and 12 cm. Drill hole spacing ranged from 14 to 52 cm with 70 percent clustering between 18 and 27 cm. A related measurement was the distance between the end drill holes and the edge of the boulder. These measurements ranged from 14 to 47 cm with the majority (76 percent) ranging from 16 to 32 cm.

Figure 81. Boulder # 33 at the Ewen Millstone Quarry. Note the vertical drill holes along the upper surface of the stone. Photograph (facing north) taken on April 25, 1990.

Some of the boulders may actually be millstone preforms. These include boulders # 1 (80 × 115 × 55 cm), # 2 (120 × 125 × 55 cm), # 3 (115 × 120 × 35 cm), # 4 (120 × 140 × 55 cm) (Figure 82), # 5 (110 × 110 × 70+ cm), # 6 (132 × 135 × 30+ cm) (Figure 83), # 9 (95 × 135 × 50 cm), # 10 (78 × 100 × 53+ cm), # 17 (85 × 105 × 35 cm), # 19 (100 × 130 ×

Left: Figure 82. Boulder # 4 at the Ewen Millstone Quarry. This is probably a millstone preform. Photograph (facing northwest) taken on April 25, 1990. *Right:* Figure 83. Boulder # 6 at the Ewen Millstone Quarry. This is probably a millstone preform. Photograph (facing east) taken on April 25, 1990.

75 cm), # 20 (105 × 125 × 35+ cm), # 22 (100 × 126 × 70 cm), # 33 (130 × 70 × 60 cm), # 35 (140 × 180 × 35 cm), # 37 (120 × 150 × 30 cm), # 39 (105 × 180 × 52 cm), and # 40 (95 × 170 × 90 cm). These reassessments are based on information contained on recording forms and in the field sketches. Boulder # 39 also has linear tool marks that are probably associated with the laying out of a leveling cross. Five tool marks are oriented east-west and two marks are oriented north-south. The east-west marks would have laid out a trough about 20 cm wide and 57 cm long.

Table 12. Drilled Boulder Measurements at the Ewen Quarry

Specimen No.	Length (cm)	Width (cm)	Thickness (cm)	No. of Holes
01	115	80	55	2
02	125	120	55	1
03	120	115	35	1
04	140	120	55	0
05	110	110	70+	5
06	135	132	30+	0
07	60	30	28	1
08	135	115	60	0
09	135	95	50	2
10	100	78	53+	2
11	58	43	60	3
12	75	35	55	2
13	50	47	27	1
14	45	44	25	2
15	65	60	50	2
16	85	60	30+	1
17	105	85	35	2
18	92	36	31	2
19	130	100	75	5
20	125	105	35+	3
21	115	87	50	2
22	126	100	70	1
23	100	90	40	1
24	90	62	26	1
25	70	40	38	1
26	110	85	40	1

Specimen No.	Length (cm)	Width (cm)	Thickness (cm)	No. of Holes
27	95	52	35	4
28	75	27	38	1
29	65	57	20	1
30	90	45	12	2
31	125	55	55	6
32	98	47	25	3
33	130	70	60	3
34	70+	50	25	1
35	180	140	35	1
36	165	160	40	0
37	150	120	30	7
38	93	65	50	1
39	180	105	52	0
40	170	95	90	3

+Bottom buried and maximum thickness unavailable.

Another source of data present on four boulders (Numbers 4, 8, 36, and 39) was the narrow linear tool marks. These tool marks (n=16) were very shallow (1–2 cm) and most ranged from 8 to 32 cm in length. The tool marks were frequently vertically oriented and roughly parallel (4–35 cm apart). Horizontal tool marks on Boulder # 39 ranged in length from 8 to 57 cm (8, 8, 12, 22, 23, 37, and 57 cm long), were 1 to 2 cm wide, and spaced between 4 and 10 cm apart (4, 6, 10, and 10 cm apart). Vertical tool marks on Boulder # 36 ranged in length from 8 to 57 cm (10, 13, 15, 15, 15, and 20 cm long), were 1 cm wide, and were spaced between 6 and 35 cm apart (6, 8, 10, 22, and 35 cm apart).

A final observation on the boulders was the type of break obtained from splitting the stone. Most boulders had irregular (n=11), concave (n=10), or convex (n=10) drilled faces; a smaller number (n=7) had straight faces. This suggests that it was difficult to split the stone perfectly straight.

Pilot Knob Quarry

The Pilot Knob Millstone Quarry (15Po307) is located near the western fringe of Pilot Knob. Fieldwork was conducted on March 4 and 27, 1987; March 16 and April 8, April 13, 1988. Quarrying remains were present in two contiguous areas. Millstones were primarily scattered along (and in) an unnamed tributary of Brush Creek (Stream A) that is oriented north-south (Figure 84). The steep slope flanking the eastern stream bank contains drilled boulders. The second area (Figure 85) with quarrying remains is a prominent north-south oriented bench (900–940 feet above MSL) overlooking Stream A. Millstones and boulders with drill holes are scattered along the bench. The bench is dissected by several shallow intermittent wet water streams (designated Stream B, C, D, E, F, and G) draining the western slope of Pilot Knob. Quarry remains are clustered at the intersection of Streams A and C and also along the upper reaches of Stream E. Streams B, D, and G have sparse quarrying remains while no quarrying remains were observed along Stream F. About 24 m east of the cluster on Stream E is a small pit 2 m in diameter and 1 m deep. Two meters north of the small pit is a small bench cut into the slope that is 30 m long (north-south) and 5 m wide (east-west). Approximately 17 m east of the small pit is another bench upslope that measures 30 m long (north-south) and 5 to 6 m wide (east-west). In the vicinity of this upper bench is a larger pit that measures 3.5 × 4.3 m and is 50 cm deep.

Figure 84. Segment of Stream A at the Pilot Knob Millstone Quarry. Photograph (facing east) taken on April 13, 1989.

Figure 85. Main Ridge at the Pilot Knob Millstone Quarry. Photograph (facing west) taken on March 27, 1987.

The quarry extends over an area approximately 320 m north-south and 100 m east-west (3.2 ha). Twelve millstones in various stages of completion and 18 boulders with drill holes were documented. Local oral history indicates that a wealthy farmer removed a number of millstones from the quarry during the 1930s. One informant related a story concerning a local man who was employed to drag millstones by mule to the road, where they were loaded onto trucks. Through extended research, a large millstone collection containing specimens reportedly from the Pilot Knob Quarry was located and documented. The current owners wish to remain anonymous. Several of the millstones in the collection were made from the fine pebble conglomerate which occurs at Pilot Knob (see Appendix G).

The typical conglomerate worked at this quarry can be characterized as follows: a light gray to tan sandstone containing rounded quartz pebbles (mostly white, some yellow and brown). Most of the pebbles are under 5 mm (some less than 1 mm) with a few ranging up to 3 cm in diameter. Since the pebble size is so small in comparison to other known quarries in the vicinity, it is easy to identify those millstones made at Pilot Knob.

Two additional millstones were documented north of the quarry along the main trail in the Spencer-Morton Nature Preserve. Larry Meadows and friends moved these millstones there from the area where Brush Creek and the unnamed tributary leading to the millstone quarry intersect. The stones were moved into the nature preserve for protection with permission from the landowners. One of these millstones contains an inset brass plaque with information about the nature preserve. Few quarry remains can be identified beyond unfinished millstones and boulders with drill holes. These include broken rock (of various sizes) resulting from millstone shaping, some small benches, and shallow oval pits.

Millstones

Twelve millstones were documented at the Pilot Knob quarry. Table 13 provides a summary of the diameter, thickness, and eye diameter for each millstone from this quarry. The following discussion describes these millstones.

Table 13. Millstone Measurements at the Pilot Knob Quarry

Specimen No.	Diameter/Size (cm)	Thickness (cm)	Eye Diameter (cm)
01	121	43	NA
02	125	48	NA
03	82	20	NA
04	70 × 80	20	NA
05	90	30	NA
06	110 × 115	30	NA
07	120 × 150	30	NA
08	114	30	NA
09	115 × 125	28	NA
10	125 × 150	25	NA
11	106	41	NA
12	115	38	NA

Millstone # 1 is a slab that has been rounded about two-thirds of the way around the circumference (Figure 86). The upper surface is relatively flat. The rounded portion has a diameter of about 121 cm and is 43 cm thick. Five vertical drill hole scars are present on the angular northern end of the slab that was not rounded. The three holes on the northwestern

Figure 86. Millstone # 1 at the Pilot Knob Millstone Quarry. Note the remnant of the parent boulder to the right. Photograph (facing northeast) taken on March 16, 1988.

edge are 2.5 cm in diameter, range in depth from 9 to 17 cm (9, 12, and 17 cm deep), and are 21 cm apart. The two drill holes on the northeastern edge are 3 cm in diameter, 10 cm deep, and 24 cm apart. The stone was rejected due to shaping errors on the sides. Also, one surface on the rounded area has spalled off.

Millstone # 2 is a roughly rounded preform (Figure 87). It is about 125 cm in diameter and is 48 cm thick. Nine vertical drill hole scars, one horizontal drill hole, and four horizontal linear tool marks are present. The drill holes are located around the circumference of the stone and were used to remove excess stone in order to make it round. The vertical drill holes are 2 to 3 cm in diameter (4 holes are 2 cm; 3 holes are 2.5 cm; and 2 holes are 3 cm in diameter), 9 to 26 cm long (9, 9, 10, 10, 10, 12, 16, 25, and 26 cm long), and the holes in closest proximity range between 6 and 37 cm apart (6, 9, 20, 22, 31, and 37 cm apart). The horizontal drill hole is 2.5 cm in diameter and 21 cm deep. The

Figure 87. Millstone # 2 at the Pilot Knob Millstone Quarry. Photograph (facing south) taken on March 16, 1988.

horizontal tool marks are associated with the east-west axis of a leveling cross. The parallel tool marks are 2.5 cm wide, 5 to 24 cm long (5, 10, 15, and 24 cm long), and 5 to 23 cm apart (5, 12, and 23 cm apart). This initial trough is 2 cm deep and 96 cm long. The width at the edge (west end) is 50 cm, 25 cm wide towards the center, and 9 cm wide at the east end. The south side of the specimen is 27 cm thick while the north side is 58 cm thick. It is not known whether the difference in thickness is due to undercutting or an effort to separate the millstone from the parent rock. The stone also has minor edge damage.

Millstone # 3 is a roughly rounded slab. The upper surface varies 7 to 8 cm vertically. The specimen is about 82 cm (82 × 90 cm) in diameter and is 20 cm thick. The northwest section has a projection beyond the circle (15 cm long and 35 cm wide). The millstone was rejected due to shaping errors.

Millstone # 4 is a roughly rounded slab with a relatively flat upper surface. It measures 70 × 80 cm in diameter and is 20 cm thick. The northwest section has a projection beyond the circle. The millstone was rejected due to shaping errors on the sides.

Millstone # 5 is a roughly rounded slab. The upper surface is relatively flat. The specimen is about 90 cm in diameter and is 30 cm thick. The northwest section has a projection that extends beyond the circle. The millstone was probably rejected due to shaping errors on the sides.

Millstone # 6 is a roughly oval slab with an irregular upper surface that varies 5 to 7 cm. It measures 110 × 115 cm and is 30 cm thick. The stone was abandoned due to irregular breaks during shaping of the sides and undercutting on one side.

Millstone # 7 is a large oval slab. The upper surface is fairly level. The slab is 120 × 150 cm and is 30 cm thick. A narrow strip on the western edge of the slab has been shaped to 13 cm below the upper surface. Another area (42 × 70 cm) on the northeast portion of the slab is 3 to 4 cm lower than the upper surface. Shaping errors seem to have led to the abandonment of this specimen.

Millstone # 8 is a roughly rounded slab with an irregular upper surface (varies 5 to 7 cm vertically). It measures 114 cm in diameter and is 30 cm thick. One unique feature of the specimen is the presence of a central hole that is 3 cm in diameter and 3 cm deep. This was probably the reference hole used to assist in outlining the circle of the millstone. Sixteen horizontal linear tool marks are on the upper surface of the stone. Six tool marks are oriented north-south, four are oriented east-west, five are oriented northwest-southeast, and one is oriented northeast-southwest. Tool marks range from 10 to 45 cm long (10, 10, 10, 10, 11, 12, 14, 14, 18, 20, 22, 23, 30, 32, and 45 cm long), 1 to 3 cm deep, and 1 to 4 cm wide, and are spaced 4 to 13 cm apart (4, 5, 5, 6, 7, 7, 8, 8, 11, and 13 cm apart). They appear to be associated with the initial layout of a leveling cross. The north-south axis of the cross would have been about 20 cm wide. The millstone was probably abandoned due to shaping errors.

Millstone # 9 is an oval slab with an upper surface that varies 3 to 4 cm vertically. It measures 115 × 125 cm and is 28 cm thick. The upper surface of the stone contains nine horizontal linear tool marks and four additional vertical tool marks (three on the north edge and one on the west edge). Three tool marks are oriented north-south and six are oriented east-west. Horizontal tool marks range from 12 to 50 cm long (12, 18, 18, 22, 28, 30, 40, 47, and 50 cm long), 1 to 3 cm deep, and 1 to 3 cm wide, and are spaced 8 to 12 cm apart (8, 9, 10, and 12 cm apart). They appear to be associated with the initial layout of a leveling cross. The east-west axis would have been about 21 cm wide. The vertical tool marks are 8 to 17 cm long, 2 cm wide, and 2 cm deep. The stone was abandoned because of irregular breaks during the shaping of the sides.

Millstone # 10 is a slab that has been rounded about two-thirds of the way around the circumference. The upper surface of the stone varies 2 to 10 cm vertically. The rounded portion has a diameter of about 125 cm, with an overall size of 125 × 150 cm, and is 25 cm thick. Major irregular breaks occurred during the shaping process on the western and eastern sides of the stone.

Millstone # 11 is an advanced stage preform (Figure 88). It was originally located west of Stream A at its confluence with Brush Creek and moved to the nature preserve for its protection. It has been rounded and has a complete leveling trough on the upper surface. The stone is 106 cm in diameter and is 41 cm thick. The north-south axis of the leveling cross extends across the entire stone, is 25 to 28 cm wide, and is 3 cm deep. The east-west axis of the leveling cross also extends across the entire surface, is 26 to 40 cm wide (25 to 26 cm wide most of the distance), and is 3 cm deep. The upper surface of the stone

Figure 88. Millstone # 11 at the Pilot Knob Millstone Quarry. Note the leveling cross on the upper surface. Photograph (facing west) taken on April 13, 1988.

contains six horizontal linear tool marks and about 25 additional vertical tool marks. Five of the horizontal tool marks are in the northwest quadrant and one tool mark is in the northeast quadrant. The three smaller marks were not measured. The two remaining horizontal tool marks are from 15 to 26 cm long, 2 cm deep, and 2 cm wide. They appear to be associated with the leveling of the northwest and southeast quadrants of the stone on either side of the north-south arm of the leveling cross. The vertical tool marks are around the outer edges of the millstone and are associated with the rounding of the sides. In general, these vertical tool marks are 2 cm wide, 2 cm deep, and up to 25 cm in length; most marks are 8 to 20 cm apart. The only visible flaw is an area of undercutting on the south side of the stone.

Millstone # 12 is a roughly shaped slab that is relatively flat on one side. The opposite side varies 5 to 6 cm in elevation. It was originally located north of the confluence of Brush Creek and Stream A, and was moved to the nature preserve for its protection. The stone's diameter is about 115 cm and it is 38 cm thick. Currently, the stone is set on edge with a brass plaque has been set in the center of the stone providing information on the Spencer-Morton Preserve. One vertical drill hole 3.5 cm in diameter and 14 cm deep is present on the side. The stone appears to have been rejected because of irregular breaks during the shaping of the sides.

The 12 millstones from the Pilot Knob Millstone Quarry represent several stages in the manufacturing sequence. These stages include advanced preforms (n=1) and oval or rounded preforms (n=11) but no nearly complete millstones (n=0) or rectangular preforms (n=0). They range in diameter from 70 cm to 1.25 m but most rejected millstones cluster between 90 cm and 1.25 m. In terms of thickness, they vary from 20 cm to 48 cm with most specimens clustering between 25 cm and 43 cm. Millstones were rejected for a variety of reasons including irregular breaks (n=11), undercutting (n=2), and edge damage (n=1). Sometimes two or three flaws occurred on the same specimen.

Six additional millstones were documented that may have been manufactured at or near the Pilot Knob millstone quarry (see Appendix G). Three of these are unfinished millstones in a collection in Bourbon County that may have been removed from the quarry in the 1930s. They include Millstone # 18 which is 1.45 m in diameter, 30+ cm thick, and has a round eye 23 cm in diameter; Millstone # 19 which is 1.32 m in diameter, 30+ cm thick, and has a round eye 22 cm in diameter; and Millstone # 21 which is 135 cm in diameter, 30+ cm thick, and has a round eye 26 cm in diameter. Two millstones in the collection of the Red River Historical Society's museum were found in the general vicinity, southwest of Pilot Knob. They include Millstone # 3 which is 1.06 m in diameter, 30 cm thick, and has a round eye 18 cm in diameter and Millstone # 8 which is 1.41 m in diameter, 33 cm thick, and has a square eye measuring 22 × 22 cm.

Other noteworthy features of the Pilot Knob millstones include eyes and leveling crosses. One complete leveling cross and two single arm crosses were encountered on the millstones. The crosses ranged in width from 9 to 28 cm (most 21–28 cm) and were 2–3 cm deep. No millstones containing eyes were found at the quarry. However, most of the complete specimens that would have contained eyes were probably removed from the site in the 1930s.

Manufacturing evidence consists of linear tool marks and drill holes. Four millstones (Numbers 2, 8, 9, and 11) had linear tool marks and three millstones (Numbers 1, 2, and 12) had drill holes. The horizontal linear tool marks (n=35) ranged in number from four to 16. Available size ranges for the horizontal marks were 5 to 50 cm long (most 10–30 cm long), 1 to 4 cm wide, 1 to 3 cm deep, and 4–23 cm apart (most 4–12 cm apart). Vertical tool marks (n=29) occurred in groups of four to 25 and were 8 to 25 cm wide, 2 cm deep, and 8 to 20 cm apart. The 16 drill holes were 2.5 to 3.5 cm in diameter, ranged in depth between 9 and 26 cm, and were 6–37 cm apart (most 20–24 cm).

Drilled Boulders

At the Pilot Knob Quarry, 16 conglomerate boulders and 2 sandstone boulders with drill holes were recorded (Table 14). In terms of size, the conglomerate boulders have the following ranges: length: 70 cm–1.70 m, width: 30 cm–1.20 m, and thickness: 30–80 cm. While there was considerable range in boulder size, the majority fell within the following range: length: 70 cm–1.55 m, width: 49–85 cm, and thickness: 30–65 cm. Thirty-five drill holes were documented (Figure 89), which can be further subdivided as 33 vertical and 2 horizontal. The drill holes were of three different diameters: 3 cm (n=15), 3.5 cm (n=14), and 4 cm (n=6). All the holes were split along their long axis. Drill hole depth ranged as follows: 3 cm diameter (7–22 cm deep) 3.5 cm diameter (9–35 cm deep) and 4 cm diameter (10–14 cm deep). However, the drill holes clustered in depth as follows: 3 cm diameter, 9–12 cm deep; 3.5 cm diameter, 10–13 cm deep; and 4 cm diameter, 13–14 cm deep. The spacing of the drill holes ranged from 16 to 58 cm with most (65 percent) clustering between 21 and 30 cm. The distance between the end drill holes and the edge of the boulders ranged from 6 to 58 cm with most (58 percent) clustering between 17 and 30 cm apart. Four fracture types were produced by these drill holes: straight (n=7), concave (n=5), irregular (n=4), and convex (n=2).

Some of the boulders may actually be millstone preforms. These include boulders # 5 (92 × 135 × 40 cm), # 7 (85+ × 135 × 50 cm), # 9 (85 × 170 × 45 cm), # 10 (120 × 130 × 40+ cm), and # 13 (80 × 150 × 80 cm). These reassessments are based on information contained on the recording forms and in the field sketches.

Tool marks were observed on three boulders. These marks (n=8) were shallow (1 cm deep) linear depressions. They range in length from 10 to 25 cm and are approximately 2 cm wide. Most have a vertical orientation and are parallel in orientation.

On the northern edge of the quarry, two drilled sandstone boulders (Table 14, specimens 17 and 18) were located. It is not known whether they represent sandstone portions of larger boulders containing conglomerate or were utilized for large grindstones. They have the following size ranges: length: 85 cm–1.58 m, width: 65 cm–1.23 m, and thickness: 34–38 cm. Altogether, eight vertical drill holes of two sizes were present. Six of the drill holes were 3.5 cm in diameter and ranged in depth from 11 to 16 cm. The remaining two drill holes were 4 cm in diameter and ranged in depth from 12 to 15 cm. All eight drill holes were split along their long axis. Drill hole spacing ranged from 14 to 42 cm with most ranging between 14 and 21 cm. The distance between the end drill holes and the edge of the boulders ranged between 9 and 24 cm. One boulder had two irregular fractures and the other had a concave fracture.

Figure 89. Boulder # 11 at the Pilot Knob Millstone Quarry. Note the vertical drill holes along the upper surface of the stone. Photograph (facing west) taken on April 8, 1988.

Boulder 18 also had three shallow vertical tool marks, which ranged in length from 13 to 18 cm and in width from 1.5 to 2 cm.

Table 14. Drilled Boulder Measurements at the Pilot Knob Quarry

Specimen No.	Length (cm)	Width (cm)	Thickness (cm)	No. of Holes
01	115	55	52+	4
02	70	62	48	3
03	118	63	40+	3
04	76	49	32	2
05	135	92	40	4
06	70	55	65	2
07	135	85+	50	1
08	140	40+	30	3
09	170	85	45	2
10	130	120	40+	4
11	95	60	46	2
12	155	70	30+	3
13	150	80	80	1
14	160	70	60	2
15	170	30	40	2
16	113	52	55	3
17	158	123	34	6
18	85	65+	38	2

+Bottom buried and maximum thickness unavailable.

7

Comparisons Among the Powell County Quarries

This chapter compares the data for the six Powell County millstone quarries. First, it looks at the data on millstone sizes, leveling crosses, eyes, drill holes, and reduction stages represented. The next section examines boulders with drill holes, comparing their dimensions and the diameters, depths, and spacing of drill holes. Subsequent sections deal with shaping debris, quarry excavations, recovered artifacts, tool marks, and reasons why millstones were rejected.

Two of these millstone quarries are located near the crest of a knob. They exploited in situ stone as well as boulders scattered on the slopes below. The remaining four quarries exploited conglomerate boulders located on ridge tops, streambeds, and adjacent areas. The quarries range in size (Table 15) from about 1 ha to 14 ha (2.5 to 35 acres). Quarry byproducts are found on ridge tops, slopes, and within streambeds. The raw material used for the millstones is conglomeratic sandstone which ranges from a light gray to light tan and contains numerous rounded quartz pebbles (mostly white with some yellow, brown, and pink). Most of the quartz pebbles are less than 2 cm in diameter, but occasionally pebbles up to 5 cm in diameter are present. Chert, sandstone, and fossil inclusions are also present in some of the conglomerate. The archaeological evidence associated with these quarries includes millstones in various stages of completion, boulders with drill holes, shaping debris, quarry excavations, tool marks on boulders and millstones, and historic artifacts such as iron tool fragments.

Table 15. Numbers of Documented Millstones, Boulders, and Sizes of the Six Powell County Millstone Quarries

Quarry	Number of Millstones	Number of Boulders	Quarry Size
McGuire	38	66	4.2 ha
Baker	28	55	14.0 ha
Toler	15	20	6.2 ha
Ware	33	30	5.9 ha
Ewen	5	40	1.0 ha
Pilot Knob	12	18	3.2 ha
Totals	131	229	34.5 ha

Millstones

Measurements were recorded for 131 millstones documented at the six quarries (Table 16). As a whole, the millstones range in size from 50 cm to 1.6 m (20 to 64 inches) in diameter, with most specimens clustering between 80 cm (32 inches) and 1.15 m (46 inches). These millstones range in thickness from 15 to 70 cm (6 to 28 inches), but most cluster between 25 and 45 cm (10 and 18 inches) in thickness. Twenty-four of the millstones contained central holes or eyes (see the section below on eyes).

Table 16. Size Ranges and Size Clusters for Millstones from the Six Powell County Millstone Quarries

Quarry	Diameters Ranges	Diameter Clusters	Thickness Ranges	Thickness Clusters
McGuire	80–150 cm	85–115 cm	20–70 cm	20–40 cm
Baker	75–140 cm	80–110 cm	15–60 cm	25–50 cm
Toler	50–160 cm	75–110 cm	15–45 cm	25–45 cm
Ware	70–136 cm	75–115 cm	20–70 cm	25–50 cm
Ewen	88–117 cm	88–117 cm	28–50 cm	40–45 cm
Pilot Knob	70–125 cm	90–125 cm	20–48 cm	25–43 cm

There are some variations in the sizes of the millstones produced at the quarries. The 38 millstones at the McGuire Quarry range in size from 80 cm to 1.5 m in diameter, with most specimens clustering between 85 cm and 1.15 m. These millstones range in thickness from 20 to 70 cm, but most cluster between 20 and 40 cm. At the Baker Quarry, the 28 millstones range in size from 75 cm to 1.4 m in diameter, with most specimens clustering between 80 cm and 1.10 m. These millstones range in thickness from 15 to 60 cm, but most cluster between 25 and 50 cm. The 15 abandoned millstones at the Toler Quarry range in size from 50 cm to 1.6 m in diameter, with most specimens clustering between 75 cm and 1.10 m. These millstones range in thickness from 15 to 45 cm, but most cluster between 25 and 45 cm. The 33 millstones found at the Ware Quarry range in size from 70 cm to 1.36 m in diameter, with most specimens clustering between 75 cm and 1.15 m. These millstones range in thickness from 20 to 70 cm, but most cluster between 25 and 50 cm. The five millstones documented at the Ewen Quarry range in size from 88 cm to 1.17 m in diameter, with most specimens clustering between 88 cm and 1.17 m. These millstones range in thickness from 28 to 50 cm, but most cluster between 40 and 45 cm. Finally, the 12 rejected millstones discovered at the Pilot Knob Quarry range in size from 70 cm to 1.25 m in diameter, with most specimens clustering between 90 cm and 1.25 m. These millstones range in thickness from 20 to 48 cm, but most cluster between 25 and 43 cm.

The five millstones found near Pilot Knob and others thought to be removed from Pilot Knob are slightly larger as a whole (Appendix G). These millstones range in size from 1.06 m to 1.45 m in diameter (1.06, 1.32, 1.35, 1.41, and 1.45 m in diameter). These millstones are all approximately 30 cm in thickness. These specimens suggest that the Pilot Knob Quarry was producing larger millstones than indicated by remaining rejects at the quarry.

Millstones cluster within similar size ranges for all six quarries. However, it appears that the two quarries that exploited in situ bedrock were producing slightly larger millstones. Bedrock deposits allowed the millstone makers to obtain larger slabs while available boulders at the other quarries restricted millstone sizes. Looking at the quarries as a whole, several millstone sizes are more common than other sizes. These sizes (diameters) in order of their frequency are: 90 cm (36 inches), 1.10 m (44 inches), 80 cm (32 inches), 96 cm (38.4 inches), and 1.15 m (46 inches). Several sizes of Red River millstones were mentioned in newspaper ads published in 1799 and 1818 (Kentucky Gazette 1799b; Kentucky Reporter 1818). The advertised sizes were 3 feet (36 inches), 3 feet 6 inches (42 inches), 3 feet 8 inches (44 inches),

3 feet 10 inches (46 inches), 4 feet (48 inches), and 5 feet (60 inches). Thus, three of the advertised sizes correspond to the abandoned sizes found at the quarries.

Millstones were encountered in various degrees of completeness at the six millstone quarries (Table 17). The millstones from the McGuire Quarry represent several stages in the manufacturing sequence. These stages include nearly complete millstones (n=8), advanced preforms (n=3), oval or rounded preforms (n=23), and rectangular preforms (n=4). At the Baker Quarry, these stages include nearly complete millstones (n=3), advanced preforms (n=2), oval or rounded preforms (n=19), and rectangular preforms (n=4). Rejected millstones at the Toler Quarry include nearly complete millstones (n=1), advanced preforms (n=7), and oval or rounded preforms (n=7). Abandoned specimens at the Ware Quarry include advanced preforms (n=6), oval or rounded preforms (n=20), and rectangular preforms (n=7). The Ewen Quarry contained nearly complete millstones (n=1), advanced preforms (n=1), oval or rounded preforms (n=2), and rectangular preforms (n=1). At the Pilot Knob Quarry, the only stages were advanced preforms (n=1) and oval or rounded preforms (n=11).

Table 17. Stages of Millstone Completeness at the Six Powell County Millstone Quarries

Quarry	Nearly Completed	Advanced Preform	Oval or Rounded Preform	Rectangular Preform	Totals
McGuire	8	3	23	4	38
Baker	3	2	19	4	28
Toler	1	7	7	0	15
Ware	0	6	20	7	33
Ewen	1	1	2	1	5
Pilot Knob	0	1	11	0	12
Totals	13	20	82	16	131

The oval or rounded preforms are present at all the quarries and are the most common category. Rectangular preforms were observed at only four quarries: McGuire, Baker, Ware, and Ewen. Advanced stage preforms were encountered at all six quarries but were most common at the Toler and Ware quarries. Nearly complete millstones were found at all the quarries except the Ware and Pilot Knob quarries.

An interesting group of millstones are those that were shaped from boulders and were abandoned before they were separated from the parent rock. In most cases, the tops of the stones were leveled and the millstones were partially cut into the top of the boulders. In several instances the eyes were partially cut. Examples of this type included three specimens from the McGuire Quarry (Millstones # 3, 15, and 31), two specimens from the Baker Quarry (Millstones # 2 and 14), three from the Toler Quarry (Millstones # 6, 7, and 8), and three millstones from the Ware Quarry (Millstones # 5, 10, and 23). At the McGuire Quarry, they were shaped down 10 to 22 cm below the top and had eyes cut to depths of 8 to 18 cm. The examples at the Baker Quarry were shaped down 16 to 32 cm below the top and had no eyes. The Toler Quarry examples were shaped down 16 to 22 cm below the top and had eyes cut to depths from 18 cm to all the way through the entire stone. The specimens from the Ware Quarry were shaped down 10 to 35 cm below the top and one millstone had an incomplete eye cut to depths of 2 to 6 cm. These millstones were abandoned due to edge damage, irregular breaks, and cracks.

Leveling Crosses

For the purposes of description and discussion, the leveling crosses were placed into four categories. The first category is the complete cross that still retains two perpendicular troughs across the millstone forming a "+." The second category is a T-shaped cross where half of the cross (portions of both troughs) was removed during the leveling process. Basically, the leveled area forms the top of the "T" while the remaining segment of trough forms the vertical part of the "T." The third category is the L-shaped cross where one-fourth of the millstone has been leveled, leaving an L-shaped portion of the troughs. The fourth category is the I-shaped cross which is the initial stage of leveling. At this stage, only one trough has been cut across a millstone.

Leveling crosses were present on millstones at three quarries, and a fourth quarry may have one example on a boulder (Table 18). At the McGuire Millstone Quarry no leveling crosses were encountered on the rounded millstones. However, Boulder # 21, a possible millstone preform, has a partial (T-shaped) leveling cross on the upper surface. The cross ranges from 20 to 24 cm wide and is 2 cm deep. The Baker Quarry contains four leveling crosses in different degrees of completeness including two I-shaped (Millstones # 5 and 10), one T-shaped (Millstone # 2), and one L-shaped (Millstone # 16). The cross ranges from 12 to 35 cm wide and 2 to 5 cm deep. No leveling crosses were found at the Toler Quarry. Six leveling crosses were recorded on millstones at the Ware Quarry. These included one complete cross (Millstone # 2), two I-shaped (Millstones # 11 and 20), two T-shaped (Millstones # 10 and 14), and one L-shaped (Millstone # 3). The crosses range from 8 to 20 cm wide and 3 to 6 cm deep. Boulder # 28 at the Ware Quarry also contained a single axis of a leveling cross (I-shaped) that ranged from 17 to 32 cm in width and 7 to 8 cm in depth. The Ewen Quarry contained no leveling crosses on the millstones but Boulder # 39 has some tool marks that could be associated with the laying out of a leveling cross. The Pilot Knob Quarry produced one complete cross (Millstone # 11) and two I-shaped (Millstones # 2 and 9) crosses. The cross ranges from 9 to 28 cm wide and 2 to 3 cm deep. As a whole, the 13 leveling crosses on millstones were 8–35 cm wide (most 12–28 cm wide) and 2–6 cm deep (most 5 cm deep).

It is interesting that no leveling crosses were found at the McGuire and Toler quarries. These are the only two quarries that exploited in situ conglomerate deposits. There may be several explanations for the absence of leveling crosses at these quarries. First, since slabs were quarried from bedrock, the upper surfaces of slabs were probably more level than the tops of boulders. Thus, leveling crosses were not necessary to achieve a level surface. Second, it may be that leveling crosses were used but no millstones were abandoned in that stage. Third, perhaps leveling crosses were used before or after the McGuire and Toler quarries were in use. Fourth, there may some abandoned millstones with leveling crosses that went undetected because of vegetation or soil covering them.

Observations on the surviving leveling crosses (Figure 90) show the stages of their use. First, a level trough was cut across one axis of a millstone. Logic would suggest that the depth of the cross was slightly lower than the lowest surface on the preform. Next, a second trough was cut across the millstone at a perpendicular angle. It is assumed that a simple level with a bubble was used to ensure that the troughs were level across both axes. Four quadrants were created when the cross was cut. The cross troughs provided level sighting lines across both axes of the millstone so that the worker could use these as reference points. Each quadrant was removed individually until the entire surface of the millstone was leveled. A paint staff was probably used to identify the high spots, which were pounded down with a bush

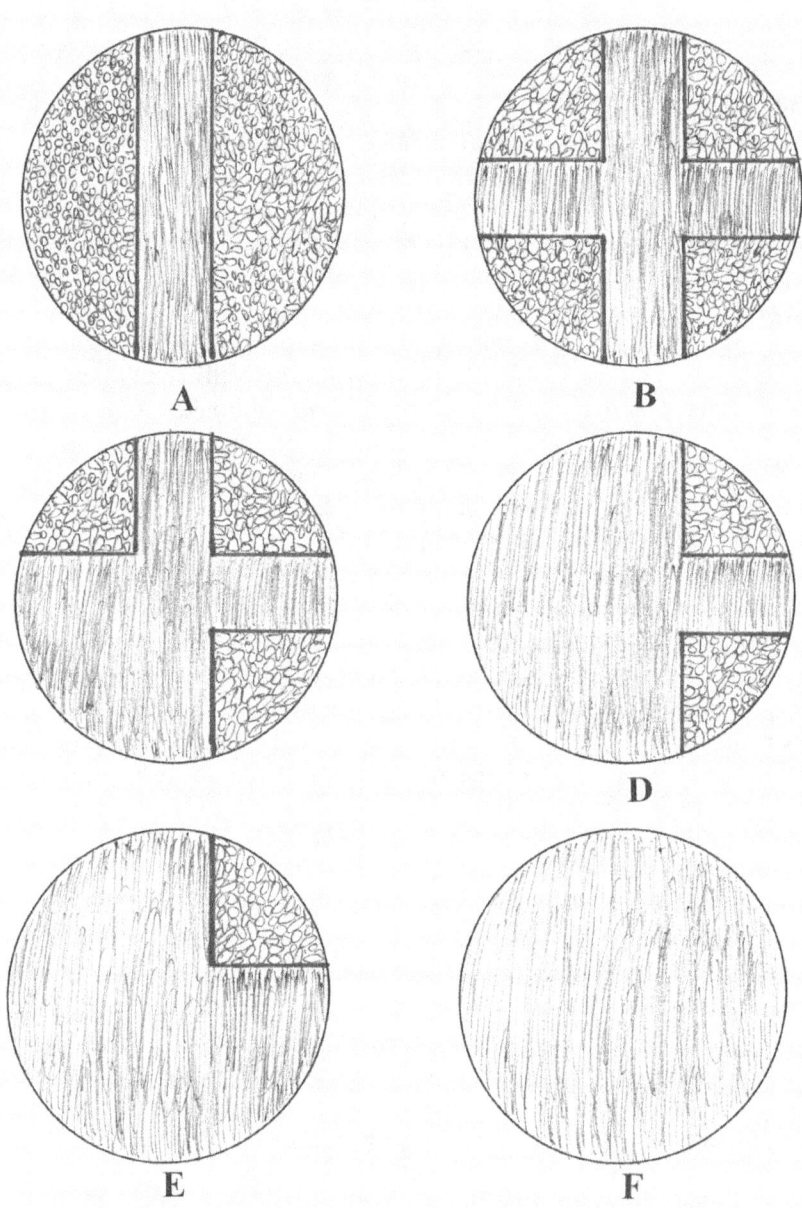

Figure 90. Drawings of the stages of cutting a leveling cross and leveling the upper surface of a Millstone: A, initial trough; B, complete cross; C, first quadrant removed; D, second quadrant removed; E, third quadrant removed; F, fourth quadrant removed.

hammer. Rejected millstones suggest that irregular breaks sometimes occurred in the stone when some of the quadrants were being removed. These unexpected fractures resulted in the millstones' being abandoned at whatever stage the break occurred. Thus, the leveling crosses discussed above occurred in different shapes depending upon their stage completion upon abandonment.

Table 18. Data on Leveling Crosses at the Six Powell County Millstone Quarries

Quarry	Complete Cross	T or L Cross	One Arm of Cross	Range of Widths	Range of Depths
McGuire	0	0	0	NA	NA
Baker	0	2	2	12–35 cm (13–15 cm)*	2–5 cm (5 cm)
Toler	0	0	0	NA	NA
Ware	1	3	2	8–20 cm (12–20 cm)	3–6 cm (5 cm)
Ewen	0	0	0	NA	NA
Pilot Knob	1	0	2	9–28 cm (21–28 cm)	2–3 cm
Totals/Ranges	2	5	6	8–35 cm (12–28 cm)	2–6 cm (5 cm)

*Numbers in parentheses indicate the most frequent size ranges.

Cutting Eyes

Twenty-four of the millstones contained central holes or eyes (Table 19). The eyes ranged from 17 to 23.5 cm (6.8 to 9.4 inches) in diameter with most eyes clustering between 16 and 20 cm (6.8 and 8 inches) in diameter. All of the eyes were round. However, one exception is a millstone in the collection of the Red River Historical Society's museum (Millstone # 8 in Appendix G). This millstone was found in the general vicinity of the Pilot Knob Millstone Quarry and has a square eye (22 × 22 cm). Historical accounts suggest that eyes were first cut round and then squared later if so desired. Since the millstones left at the quarries were rejected, final details such as squaring the eyes for bedstones would not have been done in most instances.

There are some minor variations in the sizes and completeness of the eyes at the millstone quarries. Nine millstones (Numbers 1–4, 13–15, 27, and 31) from the McGuire Millstone Quarry contained eyes. Five of the eyes were cut all the way through the millstones while the remaining four eyes were cut 8 to 22 cm deep (8, 9–12, 18, and 22 cm deep). The diameters of the eyes ranged from 17 to 22 cm with six examples between 17 and 18 cm (17, 17, 17, 17.5, 18, and 18 cm in diameter) and three examples between 20 and 22 cm (20, 21, and 22 cm in diameter). At the Baker Quarry, two millstones (Numbers 1 and 19) contained eyes. These eyes were cut all the way through the millstones and were both 17 cm in diameter. The Toler Quarry had eight millstones (Numbers 1–8) that contained eyes. Four of the eyes were cut all the way through the millstones while the remaining four eyes were cut 11 to 18 cm deep (11, 15, 17–18, and 18 cm deep). The diameters of the eyes ranged from 14 to 24 cm with six examples between 16–18 cm (16, 16, 16–18, 17, 17, and 17 cm in diameter) and two remaining examples were 14 and 24 cm in diameter. Three millstones from the Ware Quarry (Numbers 5, 27, and 29) contained eyes. These incomplete eyes ranged in depth between 2 and 14 cm. The diameters of the eyes ranged from 20 to 23.5 cm (20, 20, and 23.5 cm in diameter). The Ewen Quarry had two millstones (Numbers 1 and 2) that contained eyes. One eye was cut all the way through the millstone while the other was 8 cm deep. One eye was 17 cm in diameter and the other was 20 cm in diameter. While no millstones left at the Pilot Knob Quarry had eyes, several millstones from the vicinity had eyes (Appendix G). These millstones included four round eyes ranging from 18 to 26 cm in diameter (18, 22, 23, and 26 in diameter) and one square eye measuring 22 × 22 cm.

Table 19. The Frequency, Size, and Completeness of Eyes in Millstones at the Six Powell County Millstone Quarries

Quarry	Millstones With Eyes	Range of Eye Sizes	Clusters of Eye Sizes	Complete Eyes	Incomplete Eyes
McGuire	9	17–22 cm	17–18 cm	5	4
Baker	2	17 cm	17 cm	2	0
Toler	8	14–24 cm	16–18 cm	4	4
Ware	3	20–23.5 cm	20 cm	0	3
Ewen	2	17–20 cm	17–20 cm	1	1
Pilot Knob	0	NA	NA	0	0

Drill Holes on Millstones

Several millstones contained drill hole scars (Table 20). These holes were associated with the removal of excess stone and the separation of the millstones from parent boulders. Forty-four drill holes were documented. The number of holes ranged from one to ten per millstone. Drill hole diameters ranged from 2.5 to 3.5 cm in diameter with most holes being between 3 and 3.5 cm in diameter. Hole depths ranged from 4 to 27 cm with most holes clustering between 9 and 12 cm deep. The spacing between the holes ranged from 6 to 40 cm with most distances ranging between 20 and 27 cm. Undoubtedly, the thickness of each millstone was a factor in determining the diameter and spacing of the drill holes.

Drill holes were present on some of the millstones at all six quarries. At the McGuire Quarry, three millstones (Numbers 6, 7, and 11) had drill holes. Seven drill holes were 3 cm in diameter and one drill hole was 4 cm in diameter. They ranged in depth between 4 and 16 cm with spacing between 20 and 40 cm. Four millstones (Numbers 3, 17, 23, and 26) at the Baker Quarry produced six drill holes. Four drill holes were 3 cm in diameter and two drill holes were 3.5 cm in diameter. They ranged in depth between 7 and 27 cm (most were 7–10 cm) and ranged in space between 26 and 33 cm. The Toler Quarry had three millstones (Numbers 6, 8, and 9) with five drill holes. The drill holes were 2 cm (n=1), 3 cm (n=2), and 3.5 cm (n=2) in diameter. They ranged in depth between 8 and 18 cm and ranged in spacing between 18 and 21 cm. At the Ware Quarry two millstones (Numbers 16 and 22) produced two drill holes. The drill holes were 3 cm (n=1) and 3.5 cm (n=1) in diameter. They ranged in depth between 6 and 12 cm. The Ewen Quarry had three millstones (Numbers 3, 4, and 5) with a total of seven drill holes. The drill holes were 3 cm (n=3) and 3.5 cm (n=4) in diameter. They ranged in depth between 9 and 13 cm and ranged in space between 22 and 32 cm. Finally, the Pilot Knob Quarry had three millstones (Numbers 1, 2, and 12) with 16 drill holes. The drill holes were 2 cm (n=4), 2.5 cm (n=7), 3 cm (n=4), and 3.5 cm (n=1) in diameter. They ranged in depth between 9 and 26 cm and ranged in spacing between 6 and 37 cm (most were 20 to 27 cm).

Table 20. Drill Holes on Millstones at the Six Powell County Millstone Quarries

Quarry	No. of Drill Holes	Drill Holes Per Stone	Hole Diameters	Hole Depths	Hole Spacing
McGuire	8	1–4	3–4 cm (3 cm)*	4–25 cm (10–16 cm)	20–40 cm

Quarry	No. of Drill Holes	Drill Holes Per Stone	Hole Diameters	Hole Depths	Hole Spacing
Baker	6	1–2	3–3.5 cm (3 cm)	7–27 cm (7–10 cm)	26–33 cm
Toler	5	1–2	2–3.5 cm	8–18 cm	18–21 cm
Ware	2	1	3–3.5 cm	6–12 cm	NA
Ewen	7	1–4	3–3.5 cm	9–13 cm	22–32 cm
Pilot Knob	16	1–10	2.5–3.5 cm	9–26 cm	6–37 cm (20–24 cm)
Totals/Ranges	44	1–10	2.5–4 cm (3–3.5 cm)	4–27 cm (9–12 cm)	6–40 cm (20–27 cm)

*Numbers in parentheses indicate the most frequent size ranges.

Drilled Boulders

During the investigations, 229 boulders with drill holes were documented (Table 21). Holes were drilled in the boulders so that the stone could be split into desirable sizes and shapes. In terms of size, the boulders ranged between 30 cm and 2.9 m (1 to 9.6 feet) in length with most boulders clustering between 38 cm and 1.6 m (1.3 to 5.3 feet). Boulder width ranged between 20 cm and 2.4 m (8 inches to 7.9 feet) with most clustering between 30 cm and 1.1 m (1 and 3.6 feet) in width. Boulder thickness ranged between 15 cm and 1.75 m (6 inches to 5.8 feet) with most clustering between 25 and 80 cm (10 to 32 inches) in thickness. The boulders usually contained between one and six drill holes with most boulders containing between two and four drill holes. Some very large boulders contained between 14 and 19 drill holes each. No attempt was made to document those boulders lacking drill holes.

Stones classified as boulders probably represent several types of quarrying remains. Some specimens are probably excess stone removed during the shaping of larger boulders. Other boulders may represent early stages of millstones. Still others may be slabs prepared for further reduction. Finally, some of the boulders may have been split open so that their interiors could be examined to see if the stone was of suitable quality for making millstones (correct pebble sizes and good cementing, free of large inclusions and seams, and so forth). During the early stages of the fieldwork the author was learning to recognize quarrying remains. Consequently, if he were to re-examine many of these specimens today, he might classify them differently in light of a better understanding of the millstone industry.

Drill holes (Table 22) were recorded in the following diameters: 2.5 cm (1 inch), 3 cm (1³⁄₁₆ inches), 3.5 cm (1³⁄₈ inches), 4 cm (1⁹⁄₁₆ inches), 4.5 cm (1³⁄₄ inches), and 5 cm (2 inches). The most common sizes were 3.5 cm, 3 cm and 4 cm while the 2.5, 4.5, and 5 cm holes were very rare. Drill holes varied in depth (Table 23) from 3 to 25 cm (1.25 to ca. 10 inches) with most clustering in depth between 9 and 15 cm (3.5 to 6 inches). The holes were spaced between 7 and 53 cm (2.75 to ca. 21 inches) apart with most ranging between 18 and 31 cm (7.25 to 12.25 inches) apart (Table 24). Fracture types produced (Table 25) when the stones were split along the drill holes were straight, irregular, concave, and convex (in order of frequency).

Table 21. Boulder Size Ranges and Drill Hole Frequency Ranges for Boulders at the Six Powell County Millstone Quarries

Quarry	Lengths	Widths	Thickness	Range of Holes
McGuire	30–290 cm	20–240 cm	15–140 cm	1–14
	(50–166 cm)*	(35–110 cm)	(25–60 cm)	(1–3)
Baker	44–170 cm	15–118 cm	18–85 cm	1–6
	(48–110 cm)	(35–90 cm)	(25–60 cm)	(1–3)
Toler	38–110 cm	24–85 cm	16–70 cm	1–4
				(1–2)
Ware	65–285 cm	20–220 cm	25–175 cm	1–19
	(70–145 cm)	(35–110 cm)	(32–80 cm)	(1–3)
Ewen	45–180 cm	27–160 cm	12–90 cm	1–7
	(70–135 cm)	(35–120 cm)	(25–60 cm)	(1–3)
Pilot Knob	70–170 cm	30–120 cm	30–80 cm	1–6
	(70–155 cm)	(49–85)	(30–65 cm)	(1–3)
Ranges	30–290 cm	15–240 cm	12–175 cm	1–19
	(48–166 cm)	(35–120 cm)	(25–80 cm)	(1–3)

*Numbers in parentheses indicate the most frequent size ranges.

Table 22. Frequency of Drill Hole Diameters for Boulders at the Six Powell County Millstone Quarries

Quarry	2.5 cm	3 cm	3.5 cm	4 cm	4.5 cm	5 cm	Totals
McGuire	1	26	94	18	1	0	140
Baker	2	66	25	3	0	0	96
Toler	0	13	18	5	0	0	36
Ware	0	22	38	27	0	1	88*
Ewen	0	34	43	0	0	0	77
Pilot Knob	7**	15	14	6	0	0	42
Totals	10	176	232	59	1	1	479

*Only 88 of the 90 drill holes were complete enough for measurement.
**These 2.5 cm holes may be tool marks rather than drill holes.

Shaping Debris

Fragments of conglomerate associated with the shaping of millstones are present in varying quantities across the six quarries. In some cases the shaping debris was apparently discarded where it fell. In areas where extensive millstone shaping occurred, large quantities of conglomerate fragments were generated. These fragments were either piled into low mounds or disposed of by throwing them down slope. The greatest quantities of fragments occur near the knob crests where conglomerate outcrops were exploited. Quarry activities on slopes and ridge tops and in streambeds generated smaller quantities of debris since only scattered individual boulders were being shaped.

During the fieldwork at the six millstone quarries, no attempt was made to document unmodified shaping debris. General observations were recorded on field maps concerning density and concentrations of this debris in some instances. Because of dense leaf cover and undergrowth, much of the shaping debris was only partially exposed. Time, labor, and potential research questions were not available for such a study between 1987 and 1990. As our knowledge of millstone making increases, it may be possible to extract certain types of information from these remains in the future. In those instances where small concentrations of debris are associated with the making of isolated millstones, perhaps refit studies could be conducted. Such studies would attempt to fit fragments of stone together in an effort to determine the size of the pieces being removed and the sequence in which they were removed. While studies of this type have been successfully achieved on small flint tools, the coarse nature of conglomerate may make similar studies on millstones impossible. At any rate, in the present study little can be said about these fragments of conglomerate.

Table 23. Drill Hole Depths by Drill Hole Diameters for Boulders at the Six Powell County Millstone Quarries

Quarry	2.5 cm	3 cm	3.5 cm	4 cm	4.5 cm	5 cm
McGuire	7 cm	8–14 cm (10–13 cm)*	9–32 cm (10–16 cm)	3–14 cm	10 cm	---
Baker	10 cm	3–22 cm (9–12 cm)	5–14 cm (7–14 cm)	14–15 cm	---	---
Toler	---	7–22 cm (9–12 cm)	5–17 cm (10–15 cm)	10–15 cm (10–12 cm)	---	---
Ware	---	3–16 cm (8–12 cm)	7–15 cm (9–12 cm)	4–14 cm (10–14 cm)	---	15 cm
Ewen	---	7–15 cm (8–12 cm)	8–15 cm (10–14 cm)	---	---	---
Pilot Knob	10–15 cm	7–22 cm (9–12 cm)	9–35 cm (10–13 cm)	10–14 cm (13–14 cm)	---	---

*Numbers in parentheses indicate the most frequent depths of drill holes.

Table 24. Drill Hole Numbers and Spacing for Boulders at the Six Powell County Millstone Quarries

Quarry	No. of Holes	Hole Spacing (Range)	Hole Spacing (Most Frequent)	Space From Edge* (Range)	Space From Edge (Most Frequent)
McGuire	140	7–53 cm	14–30 cm	8–54 cm	15–32 cm
Baker	96	10–46 cm	18–31 cm	10–55 cm	17–36 cm
Toler	36	16–42 cm	21–29 cm	9–46 cm	17–30 cm
Ware	90	4–47 cm	20–31 cm	8–78 cm	14–32 cm
Ewen	77	14–52 cm	18–27 cm	14–48 cm	16–32 cm
Pilot Knob	35	16–58 cm	21–30 cm	6–58 cm	17–30 cm
Total/Ranges	474	4–58 cm	14–31 cm	6–78 cm	14–36 cm

*Space from last drill hole on each end to the edges of the boulder.

Table 25. The Frequency of Fracture Types in Split Boulders at the Six Powell County Millstone Quarries

Quarry	Concave	Convex	Straight	Irregular	Totals
McGuire	25	2	28	12	67
Baker	13	5	24	11	53
Toler	2	4	6	9	21
Ware	6	5	9	8	28
Ewen	10	10	7	11	38
Pilot Knob	5	2	7	4	18
Totals	61	28	81	55	225

Quarry Excavations

Excavations undertaken to expose conglomerate deposits include shallow oval pits, linear pits, and benches (Table 26). The shallow oval pits (n=69) appear to be associated with the uncovering of individual boulders or clusters of boulders. These pits range in size from one to six meters (ca. 39 inches to 19.75 feet) in diameter with most ranging between two and three meters in diameter. Their depth ranged from 30 cm to 1.5 m (1 to 4.95 feet) but most pits were between 50 cm and 1 meter (20 to 40 inches) in depth. Sometimes portions of quarried boulders with drill holes remain in the bottoms of these pits. Occasionally, partially quarried boulders showing a sequence of slabs being removed are still present. One extremely large oval pit (17 × 18 m and 1.5 m deep) was documented near a knob crest where in situ bedrock was being exposed.

The linear pits and benches appear to be associated with the removal of overburden to expose in situ conglomerate deposits. The linear pits (n=8) range in length from 6 to 33 m (19.8 to 108.9 feet), in width from 2 to 5 m (6.6 to 16.5 feet), and in depth from 30 cm to 3 m (1 to 9.9 feet). The benches (n=9) range from 30 to 88 m (99 to 290.4 feet) in length and from 5 to 20 m (16.5 to 66 feet) in width with highwalls ranging 2 to 9 m (6.6 to 29.7 feet). Most linear pits and benches occur near the knob crest and adjacent slopes where conglomerate outcrops were being exposed. However, linear pits and benches are sometimes found on ridges.

At the McGuire Quarry, six oval pits, three linear pits, and five benches were present. The Upper Quarry contains a large bench (46 m north-south and 10 m east-west), an oval pit 8 × 12 m which ranges 2–3 m deep, and a large crescent shaped pit (2–4 m deep) which is 46 m long and varies from 2 to 5 m in width. On the southeast end of the crescent pit is a narrow (up to 6 m wide) bench about 27 m long with a rear wall 3 m high. The Middle Quarry contains two narrow benches and an irregular pit. The largest pit is 3 × 15 m and 1 to 1.5 m deep. The Lower Quarry contains pits on Ridges B and D. Ridge B contains three oval quarry pits (2 × 2 × 0.5 m; 2 × 2 × 0.5 m; and 5 × 5 × 1.5 m). Ridge D contains a large bench (20 × 80 m) and two oval quarry pits, 4 × 5 m (1 m deep) and 3 × 4 m (1.5 m deep).

A total of 31 pits were encountered at the Baker Quarry. These include six pits on Ridge A, six pits on Ridge B, and 19 pits on Ridge C. On Ridge A, the oval pits range from 2 to 6 m in diameter (2, 2, 2, 3, 6, and 6 m) and 50 cm to 1 m in depth (50, 50, 50, 50, 100, and 100 cm). The oval pits on Ridge B vary in size from 1 m in diameter up to 3 × 4 m (1, 1.5, 2, 2 × 2, 2 × 3, 3 × 4 m) across with depths ranging from 30 to 50 cm (30, 30, 50, 50, and 50 m). Sixteen oval pits and three linear pits were located on Ridge C. The oval pits ranged in size from 1 m to 4 × 5 m (1, 1, 1, 1, 1.5, 1.5, 1.5, 1.5, 2, 2 × 3, 2 × 4, 2 × 4, 2 × 5, 2 × 6, 3 × 4, and 4 × 5 m) and in depth from 30 cm to 1.5 m (30, 30, 30, 30, 30, 30, 30, 30, 30, 30, 50, 50, 50, 50 cm, 1, and 1.5 m). The three linear pits measured 1.5 × 16 m (75 cm deep), 1.5 × 5 m (30 cm deep), and 2 × 9 m (1 m deep).

Table 26. Archaeological Remains from the Six Powell County Millstone Quarries

Quarry Name	Number of Oval Pits	Number of Linear Pits	Number of Benches
McGuire	6	3	5
Baker	28	3	0
Toler	1	1	2

Quarry Name	Number of Oval Pits	Number of Linear Pits	Number of Benches
Ware	22	1	0
Ewen	10	0	0
Pilot Knob	2	0	2
Totals	69	8	9

The Toler Millstone Quarry contained two benches, one irregular pit, and one oval pit. The main bench is about 88 m long and 40 m wide. At the southern end of this bench, a large quarry depression has been excavated into the side of the knob. This depression is about 30 by 15 m with a rear wall 9 m high. A second bench on the northern end of the quarry is approximately 15 m wide and 60+ m long. A single large oval pit is about 17 × 18 m in diameter and 1 to 2.5 m deep. The irregular pit covers an area about 15 × 22 m. It has three projections and a maximum depth of 2.5 m.

At the Ware Quarry, 23 pits were located. Pits excavated to expose boulders were present on Ridge A and on the southern end of Ridge B. Five pits were noted on Ridge A and three pits had worked boulders. The pits on Ridge A had the following sizes (length, width, and depth): 2.5 × 2.5 × 0.75 m; 2.5 × 2.5 × 0.75 m; 3 × 3 × 1 m; 3.5 × 2.5 × 0.75 m; and 5 × 6 × 1.5 m. Three oval pits were located on Ridge B and one of these pits had millstone and two pits had worked boulders. These pits had the following dimensions: 1.5 × 1.5 × 0.25 m; 1.5 × 1.5 × 0.5 m; and 2 × 2 × 0.75 m. Fifteen pits were documented on Ridge C with the following dimensions: 1.5 × 1.5 × 0.3 m; 1.5 × 1.5 × 0.5 m; 1.5 × 1.5 × 0.5 m; 1.5 × 1.5 × 0.75 m; 2 × 2 × 0.3 m; 2 × 2 × 0.5 m; 2 × 2 × 0.5 m; 2.5 × 2.5 × 0.5 m; 3 × 2 × 1.5 m; 3 × 2.5 × 0.75 m; 3 × 3 × 0.75; 3.5 × 2.5 × 1 m; 5 × 4 × 1 m; 7 × 3 × 0.75 m; and 9 × 6 × 0.3 m. At least nine of these pits had worked boulder remnants remaining in their bottoms.

Ten oval pits were observed at the Ewen Quarry. Pits excavated to expose boulders were present on Ridge A and on the southern end of Ridge B. Six pits were noted on Ridge A and two of these pits contained boulders in them. The pits on Ridge A had the following sizes (length, width, and depth): 2 × 2 × 0.75 m; 3 × 3 × 0.5 m; 3 × 3 × 0.75 m; 3 × 3 × 1 m; 4 × 3 × 0.75 m; and 4 × 4 × 1 m. Four oval pits were located on Ridge B and one of these pits still had a worked boulder (Figure 77). These pits had the following dimensions: 1.5 × 1.5 × 0.3 m; 2 × 2 × 0.5 m; 3 × 3 × 1 m; and 5 × 3 1.5 m.

The eastern edge of the Pilot Knob Quarry contained two benches and two oval pits. One oval pit is 2 m in diameter and 1 m deep. A small bench cut into the slope of the ridge is 30 m long and 5 m wide. The second bench, which is further east, measures 30 m long and 5 to 6 m wide. The second pit is near this upper bench. It measures 3.5 × 4.3 m and is 50 cm deep.

Artifacts

Several iron artifacts were discovered by Wayne Webb at the Baker, McGuire, Toler, and Ware millstone quarries. Subsequently, the Red River Historical Society searched portions of the McGuire and Ware quarries. In both cases, a metal detector was used to locate artifacts. Steve Knox, Leif Meadows, and Larry G. Meadows were the key people involved in this search for tools associated with the millstone industry. These artifacts are curated at the Red River Historical Society Museum in Clay City, Kentucky. Twenty-four of the more complete artifacts in a display case were available for study. These artifacts included specimens found by both Wayne Webb and Steve Knox. The other artifacts were in storage and thus unavailable at the time of this writing. It should be noted that the Red River Historical Society's work was directed and documented by Leif Meadows who was trained as an archaeologist. These artifacts will be available to future researchers.

The author has had the opportunity to examine actual tools used in conglomerate millstone making in New York and Virginia. This experience was an important aid in identifying the tools and tool fragments recovered from the Powell County millstone quarries. The 24 artifacts examined appear to be tools and tool fragments that were either lost or discarded at the quarries. Available accounts of millstone making suggest that a blacksmith was usually present at quarries to sharpen and repair the iron tools that quickly dulled (Hockensmith and Coy 1999, 2008a, 2008b; Hockensmith and Price 1999). As previously noted, most of the recovered tool fragments were probably broken beyond repair and simply discarded. Complete tools may have been dropped or laid down and thus accidentally lost in the dense leaf cover.

The recognizable tool types recovered from the Powell County millstone quarries include points (n=3), wedges (n=10), feathers (n=8), and miscellaneous (n=3). The points were struck with a hammer to shape the millstones. Two feathers and one wedge were used in each drill hole when stone was split apart. It is unlikely that tools such as hammers, pry bars, shovels, and squares would be commonly lost or discarded at quarries.

The 24 tools and fragments are discussed in the subsequent paragraphs. The information included on the artifact tags by the Red River Historical Society Museum was included in parentheses for each tool.

Three points (Figure 91) were in the sample of tools in the Red River Historical Society's collection. Wayne Webb found the best-preserved specimen. This tool is 20.7 cm long and has a head a slightly wider than the shaft. The striking surface of the head (3.5 × 3.4 cm) is slightly curved. The last 2 cm of the distal end of the tool is tapered to a point on all four sides. The shaft tapers from 2.6 to 2.8 cm at the proximal end and from 2 to 2.4 cm at the distal end. The shaft is four-sided but all four edges have been flattened the length of the shaft. It is assumed that flattening the edges made the tool easier to hold. The second point (Reference No. 6) was recovered from a streambed in the Lower McGuire Quarry (in the vicinity of Millstones 13 and 14) by Steve Knox. This specimen is complete but has a heavy buildup of rust and corrosion in places. This tool is 22.1 cm long and has a flared head with a striking surface measuring 4.3 × 3.8 cm. The lower 2 cm of the

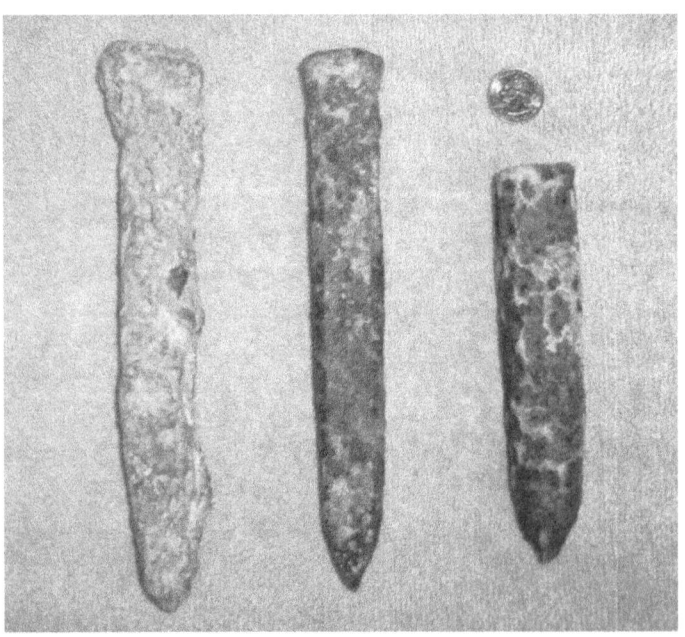

Figure 91. Three points found at the Powell County millstone quarries: Left to Right, McGuire Quarry (Reference 6); Webb (M-D); and Webb (No. V).

distal end is tapered to a point from all four sides. The shaft ranges from 3 to 2.4 cm at the proximal end to 2 to 2.3 cm at the distal end. The shaft is four sided but too corroded to see the details. The third specimen is the distal end of a broken point found by Wayne Webb. It measures 14.6 cm long and the metal is in excellent condition like the first specimen. The four-sided shaft tapers from 2.7 to 3.1 cm to 1.9 to 2.5 cm at the distal end. The last 2 cm of the shaft has been tapered to a four-sided point. The four edges of the shaft have been flattened the entire length.

Four complete wedges (Figure 92) and six wedge fragments were examined. The largest wedge (Wayne Webb, No. IV) is 11.5 cm long with a flaring head. The incomplete head is very battered and measures 5 × 3.7 cm. The shaft width varies from 3.2 cm at the proximal end to 2.5 cm at the distal end. In terms of thickness, the specimen is 2.5 cm wide at the proximal end and tapers to less than 1 cm at the distal end (battered end). The second specimen (Steve Knox, Reference No. 12) from the Ware Quarry is 10.3 cm long with a very battered head (4.6

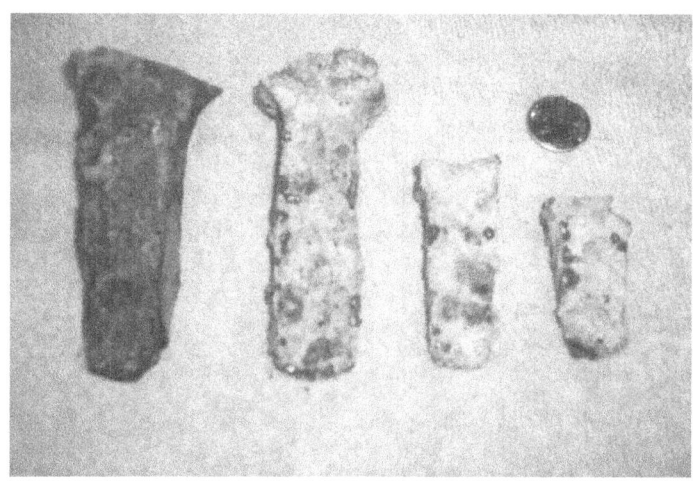

Figure 92. Four wedges found at the Powell County millstone quarries: Left to Right, Webb (No. IV); Ware Quarry (Knox, Reference 12); Ware Quarry (Knox, Reference 1); and Ware Quarry (Knox, Reference 11).

× 1.3 cm). The edges of the head have been bent away from the distal end, flaring upward. The shaft width varies from 2.6 cm at the proximal end to 2.7 cm at the distal end. In terms of thickness, the specimen is 1.7 cm wide at the proximal end and tapers to less than 6 mm at the dull distal end. The third specimen (Steve Knox, Reference No. 1) from the Ware Quarry is 6.8 cm long with a possible head (3 × 2 cm). The shaft width varies from 2.8 cm at the proximal end to 1.9 cm at the distal end. In terms of thickness, the specimen is 2 cm wide at the proximal end and tapers to less than 4 mm at the distal end. The fourth specimen (Steve Knox, Reference No. 11) from the Ware Quarry is 5.3 cm long with a head (2.9 × 1.7 cm). The shaft width varies from 3 cm at the proximal end to 2 cm at the distal end. In terms of thickness, the specimen is 1.6 cm wide at the proximal end and tapers to less than 4 mm at the distal end.

The six wedge fragments include two distal ends, one proximal end, and three shaft fragments. The largest head (Steve Knox, Reference No. 13) from the Ware Quarry is very battered. The irregular head measures 5.2 × 4.5 cm and 2.5 cm thick. The visible portion of the shaft measures 2.8 × 2.6 cm. The second head fragment (Steve Knox, Reference No. 11) from the Ware Quarry is 3.6 cm long. It is 2.9 cm wide at one end and 3.1 cm wide at the other end. The thickness ranges from 1.9 to 2.7 cm. The base measures 3 × 3 cm. The distal fragment (Steve Knox, Reference No. 2) from the Ware Quarry is 4.5 cm long. It is 2.2 cm wide at the broken end and tapers to 1.8 cm at the distal end. In terms of thickness, it is 1.5 cm thick at the broken end and tapers to 5 mm at the distal end. The first shaft fragment (Steve Knox, Reference No. 9) from the Baker Quarry is 4.4 cm long. It varies in width from 2.5 cm at the proximal end to 2 cm at the distal end. In terms of thickness, the specimen is 1.8

cm wide at the proximal end and tapers to less than 4 mm at the distal end. The second shaft fragment (Steve Knox, Reference No. 11) from the Ware Quarry is 4.8 cm long. It varies in width from 3 cm at the proximal end to 2.4 cm at the distal end. In terms of thickness, the specimen is 1.5 cm wide except where corrosion makes it 2.5 cm. The third shaft fragment (Steve Knox, Reference No. 11) from the Ware Quarry is 4.3 cm long. It varies in width from 2.9 cm at the proximal end to 2.6 cm at the distal end. In terms of thickness, the specimen is 2.6 cm wide at the proximal end to 1.7 cm at the distal end. One side is slightly concave.

Eight feathers (Figure 93) in varying degrees of completeness were analyzed. The first specimen (Wayne Webb, No. XI) is 16.5 cm long and is bent. This specimen ranges from 2 cm to 1.5 cm in width. A groove (4 cm long, ca. 4 mm wide, and ca. 4 mm deep) is located on the upper end. In terms of thickness, this feather varies from 6 to 8 mm. The last 2 cm of the distal end is chisel shaped and 2–3 mm thick. The second specimen (Steve Knox, Reference No. 4) is 10.6 cm long and is slightly bent at one end. This specimen ranges from 2 cm to 1.5 cm in width. In terms of thickness, this feather varies from 5 to 8 mm on the bent end to 1 cm on the other end. The third specimen (Wayne Webb, No. 1) is 13.6 cm long and is slightly bent on one end. This specimen is 1.8 cm wide on the bent end, 2.5 cm wide in the middle, and 2.1 cm wide at the other end. In terms of thickness, this feather varies from 2 to 4 mm on the bent end to 6 to 8 mm on the rest of the length. The fourth specimen (Wayne Webb, No. VIII) is 13.5 cm long and is slightly bent at one end. This specimen is 2 cm wide on each end and 2.3 to 2.5 cm wide elsewhere. In terms of thickness, this feather varies from 4 mm on the bent end to 6–8 mm on the rest of the length.

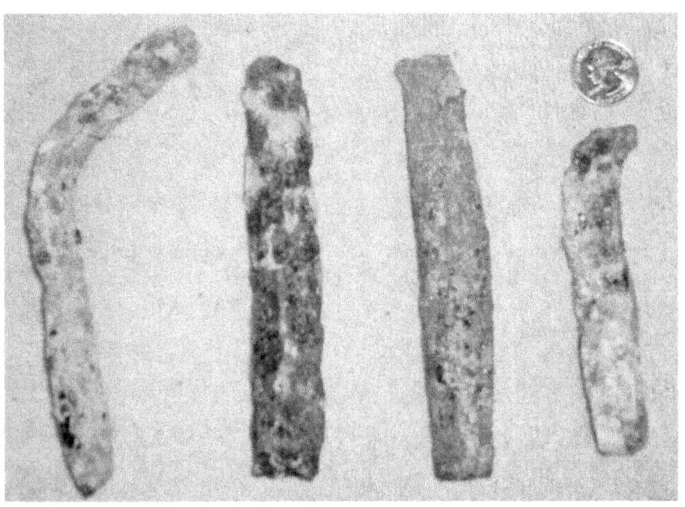

Figure 93. Four feathers found at the Powell County millstone quarries: Left to Right, Webb (No. XI); Webb (No. I); Webb (No. VIII); and Knox (Reference 4).

The fifth specimen (Steve Knox, Reference No. 10) from the Baker Quarry is 14.5 cm long. This specimen ranges from 2.2 cm to 2.7 cm in width. In terms of thickness, this feather varies from 3 mm on one end to 1.2 cm on the other end. The sixth specimen (Steve Knox, Reference No. 14) is 10 cm long. This specimen ranges from 3.2 cm to 3.3 cm in width. In terms of thickness, this specimen varies from 4 to 5 mm. The seventh specimen (Wayne Webb, No. II) is flat on one side and rounded on the other side. It is 10.6 cm long and ranges from 2.8 cm to 3.1 cm in width. In terms of thickness, this tool varies from 8 mm on one end to 1 cm on the other end. A possible feather fragment (Steve Knox, Reference No. 7) from the Baker Quarry is 3.9 cm long. This specimen ranges in width from 2.1 to 2.2 cm and is 5 to 7 mm thick.

Three miscellaneous metal fragments were examined. The first specimen (Steve Knox, Reference No. 3) from the Ware Quarry is 14.6 cm long. It is slightly wedge-shaped in profile. It is 3 cm wide on one end and narrows to 1.6 cm on the opposite end. In terms of thickness, most of the specimen is 1 cm thick but it narrows to 6 mm at one end. The second specimen

(Wayne Webb, No. IX) is an iron bar (8 cm long) that is flat on one side and slightly rounded on the other side. The width varies from 3 cm to 2.9 cm. Its thickness varies from 9 mm to 1.1 cm. The third specimen (Steve Knox, Reference No. 8) from the Baker Quarry is a small fragment 3.9 cm long. The width is 2 cm while the thickness is 7 mm.

It is expected that the tool kits used by the Powell County millstone makers would vary between the quarries exploiting in situ conglomerate (McGuire and Toler quarries) and those quarries where boulders were exploited (Baker, Ewen, Pilot Knob, and Ware). In the quarries comprised of boulders, suitable boulders were identified and simply shaped where they lay. The presence of oval pits at some of these quarries suggests that shovels were employed to uncover the buried bases of some boulders.

For the most part, it is assumed that points, various sizes of hammers, calipers, a leveling staff, feathers and wedges, and drills were the chief tools used. However, at the quarries exploiting in situ conglomerate, other tools would also be required. The removal of overburden would have required shovels, picks, axes, sleds or wagons, and possibly horse-drawn scoops. Once the conglomerate deposits were uncovered, slabs would have been quarried with sledgehammers and drills, then split with wedges and feathers. A derrick would have been necessary to lift and move the quarried slabs. Large pry bars were undoubtedly needed to raise the slabs enough to attach chains or cables to them. As in the other quarries, however, points, various sizes of hammers, calipers, a leveling staff, feathers and wedges, and drills were the chief tools used.

Tool Marks on Boulders and Millstones

Tool marks were observed on 21 boulders and 41 millstones. Most of the marks are linear depressions. The boulders contained a total of 86 tool marks (Figures 94–95). They range in width from 1 to 2 cm (⁷⁄₁₆ to ¾ of an inch), most 2 cm wide; are very shallow (1 to 2 cm);

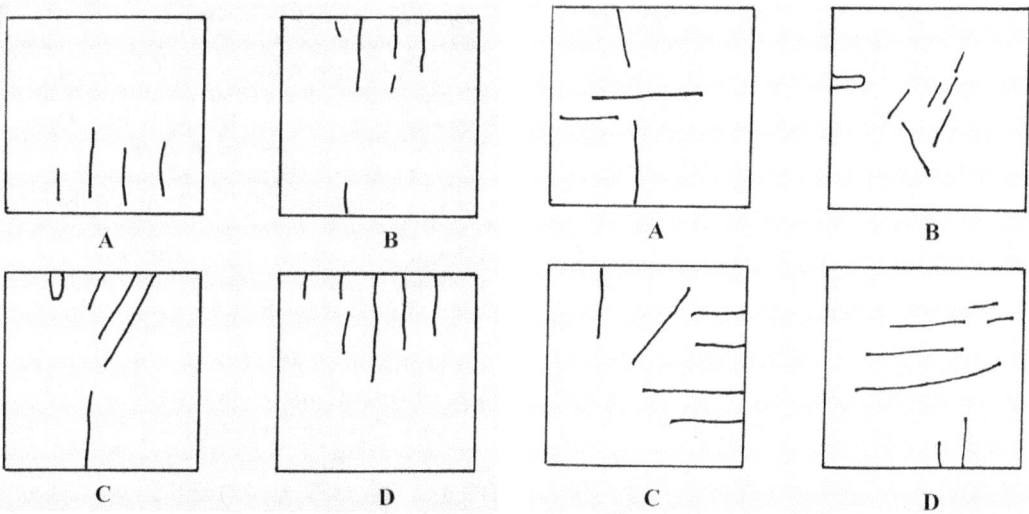

Left: Figure 94. Drawings of horizontal tool marks on millstones and boulders: A, McGuire Quarry Millstone # 24; B, Baker Quarry Millstone # 6; C, Baker Quarry Millstone # 23; and D, Baker Quarry Boulder # 5. Sketches not to scale. *Right:* Figure 95. Drawings of horizontal tool marks on millstones and boulders: A, Ware Quarry Millstone # 8; B, Ware Quarry Boulder # 2, west face; C, Ware Quarry Boulder # 29; and D, Ewen Quarry Boulder # 39. Sketches not to scale.

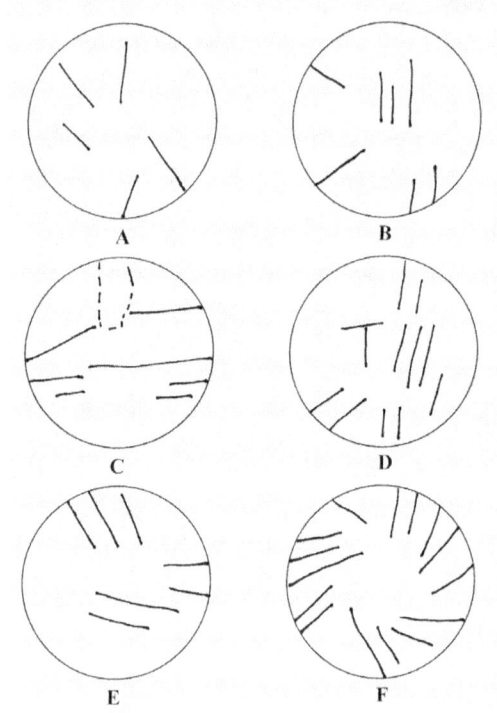

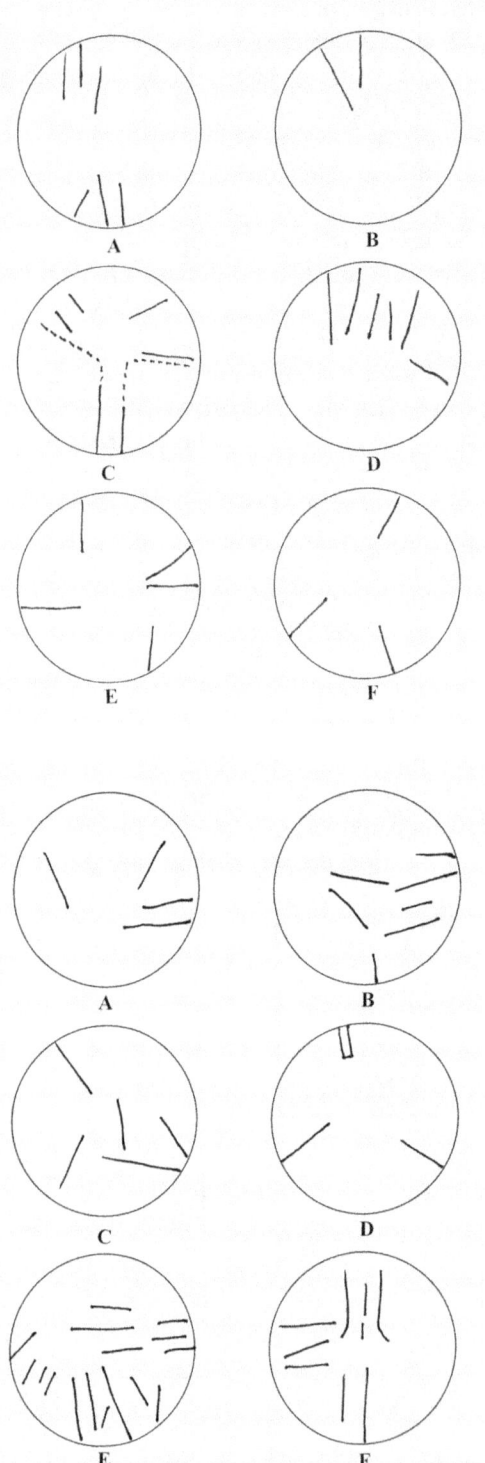

Top, left: Figure 96. Drawings of horizontal tool marks on millstones: A, McGuire Quarry Millstone # 29; B, Baker Quarry Millstone # 1; C, Baker Quarry Millstone # 5; D, Baker Quarry Millstone # 12; E, Baker Quarry Millstone # 25; and F, Baker Quarry Millstone # 26. Sketches not to scale.

Top, right: Figure 97. Drawings of horizontal tool marks on millstones: A, Toler Quarry Millstone # 9; B, Ware Quarry Millstone # 6; C, Ware Quarry Millstone # 11; D, Ware Quarry Millstone # 16; E, Ware Quarry Millstone # 17; and F, Ware Quarry Millstone # 23. Sketches not to scale.

Left: Figure 98. Drawings of horizontal tool marks on millstones: A, Ware Quarry Millstone # 24; B, Ware Quarry Millstone # 26; C, Ware Quarry Millstone # 28; D, Ewen Quarry Millstone # 4; E, Pilot Knob Quarry Millstone # 8; and F, Pilot Knob Quarry Millstone # 9. Sketches not to scale.

vary greatly in length (6 to 67 cm; 2⅜ to 26½ inches); and are spaced between 3 and 74 cm (1³⁄₁₆ to 29⅛ inches) apart, most 8 to 20 cm apart. For millstones, 187 horizontal tool marks (Figures 96–98, Table 27) and 54 vertical tool marks (Table 28) were documented. The horizontal tool marks are 1–5 cm wide (most 2 cm wide), 4 to 80 cm long (most cluster between 10 and 42 cm), 1 to 4 cm deep, and 4 to 29 cm apart (most 4 to 17 cm apart). The vertical tool marks (Figure 99) are 2–4 cm wide (most 2 cm wide), 8 to 32 cm long, 2 cm deep, and

5 to 21 cm apart. The tool marks on the upper (horizontal) surfaces appear to be chisel marks associated with the laying out of leveling crosses. Similar tool marks on vertical surfaces may be associated with the final shaping of millstone sides. One linear tool mark was recorded that was produced by a series of shallow drill holes placed in a straight line. Table 29 provides information on tool marks occurring on boulders.

At the McGuire Quarry, 12 tool marks were recorded on boulders and 35 tool marks on millstones. Tool marks (n=12), in addition to drill holes, were observed on four boulders (Numbers 21, 29, 54, and 62). These were very shallow (1–2 cm wide) linear depressions cut into the sides and tops of boulders. They range between 13 and 25 cm in length. The examples on the sides of boulders were vertical and parallel to one another. The examples on a boulder (Number 21) top appeared to be associated with the early stage of a leveling cross. Eight millstones (Numbers 4, 9, 12, 24, 26, 29, 33, and 37) at

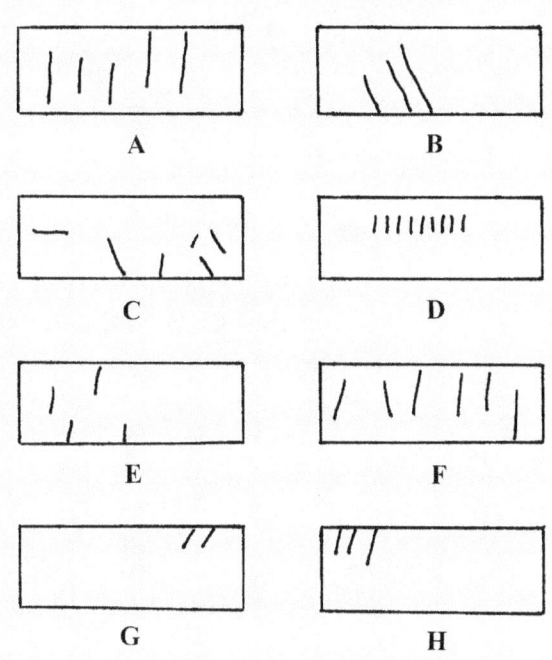

Figure 99. Drawings of vertical tool marks on millstones and boulders: A, McGuire Quarry Millstone # 9; B, McGuire Quarry Millstone # 37; C, Ware Quarry Boulder # 1, north face; D, Ware Quarry Boulder # 1, west face; E, Ewen Quarry Millstone # 3; F, Ewen Quarry Boulder # 36; G, Pilot Knob Quarry Boulder # 7; and H, Pilot Knob Quarry Boulder # 18. Sketches not to scale.

the McGuire Quarry had linear tool marks. The horizontal linear tool marks ranged in number from three to six and one specimen had a radial pattern (Millstone 4). Available size ranges for the horizontal marks were 6 to 22 cm long, 1 to 3 cm wide, and 1 to 3 cm deep. Vertical tool marks occurred in groups of three and five and ranged in length between 12 and 32 cm; they were spaced 5 to 21 cm apart.

Table 27. Data on Horizontal Tool Marks on Millstones at the Six Powell County Millstone Quarries

Quarry	No. of Horizontal Holes	Horizontal Hole Widths	Horizontal Hole Lengths	Horizontal Hole Depths	Horizontal Hole Spacing
McGuire	27	1–5 cm (2 cm)*	14–65 cm (14–25 cm)	1–4 cm	5–25 cm (5–17 cm)
Baker	39	2–5 cm (2 cm)	5–50 cm (10–35 cm)	1–2 cm	5–22 cm (5–17 cm)
Toler	9	1.5–3 cm	9–32 cm (21–32 cm)	---	26–29 cm
Ware	74	1–3 cm (2 cm)	4–80 cm (10–42 cm)	2–3 cm	4–26 cm (6–17 cm)
Ewen	3	1–2 cm	13–21 cm	1 cm	---

Quarry	No. of Horizontal Holes	Horizontal Hole Widths	Horizontal Hole Lengths	Horizontal Hole Depths	Horizontal Hole Spacing
Pilot Knob	35	1–4 cm	5–50 cm (10–30 cm)	1–3 cm	4–23 cm (4–12 cm)
Totals/Ranges	187	1–5 cm (2 cm)	4–80 cm (10–42 cm)	1–4 cm	4–29 cm (4–17 cm)

*Numbers in parentheses indicate the most frequent size ranges.

Table 28. Data on Vertical Tool Marks on Millstones at the Six Powell County Millstone Quarries

Quarry	No. of Vertical Holes	Vertical Hole Widths	Vertical Hole Lengths	Vertical Hole Depths	Vertical Hole Spacing
McGuire	8	---	12–32 cm	---	5–21 cm
Baker	6	2–4 cm (2 cm)*	8–32 cm	---	16 cm
Toler	---	---	---	---	---
Ware	6	2 cm	18 cm	---	12–14 cm
Ewen	5	---	---	---	---
Pilot Knob	29	2 cm	8–25 cm	2 cm	8–20 cm
Totals/Ranges	54	2–4 cm (2 cm)	8–32 cm	2 cm	5–21 cm

*Numbers in parentheses indicate the most frequent size ranges.

The Baker Quarry contained 13 tool marks on boulders and 45 tool marks on millstones. Tool marks (n=13) were recorded on four boulders (Numbers 5, 18, 25, and 55). Most of these are narrow (2 cm) linear depressions cut into the sides and tops of boulders. They range between 8 and 20 cm apart and are parallel. On Boulder # 5, several of these appear to be associated with a leveling cross. This specimen indicates that tool marks were formed by placing shallow drill holes along a straight line and removing the stone from between the holes. Another type of tool mark was observed on Boulder # 24. This is a 67 cm long and 3–3.5 cm wide, shallow (1–2.5 cm deep) depression on the top of the boulder. This may be an abrading area for tool sharpening. Thirteen millstones (Numbers 1, 3, 4, 5, 6, 8, 12, 18, 20, 21, 23, 25, and 26) had linear tool marks. The horizontal linear tool marks (n=39) ranged in number from one to seven. Four specimens had a radial pattern (Millstones 1, 12, 23, and 26). Available size ranges for the horizontal marks were 5 to 50 cm long (most 10–35 cm), 2 to 5 cm wide, 1 to 2 cm deep, and 5–22 cm apart (most were 5–17 cm). Vertical tool marks (n=6) occurred in groups of one and two marks. They ranged between 8 and 32 cm long, 2–4 cm wide (most 2 cm), and 16 cm apart.

At the Toler Quarry, two tool marks were recorded on boulders and nine tool marks on millstones. Two tool marks were noted on Boulder # 10. These are narrow (2 cm wide) linear shallow depressions. The tool mark on the upper surface is 8 cm long and is 19 cm from the edge of the slab. The second tool mark, on the side of the slab, is 12 cm long and 8 cm below the upper surface. One specimen is on the top of the boulder and the other is on the side. Two millstones (Numbers 2 and 9) had linear tool marks. The horizontal linear tool marks (n=9) ranged in number from four to five. Available size ranges for the horizontal marks were 9 to 32 cm long (most were 21–32 cm), 1.5 to 3 cm wide, and 26 to 29 cm apart. No vertical tool marks were observed on the millstones.

Table 29. Data on Tool Marks on Boulders at the Six Powell County Millstone Quarries

Quarry	No. of Holes	Hole Widths	Hole Lengths	Hole Depths	Hole Spacing
McGuire	12	1–2 cm	13–25 cm	1–2 cm	---
Baker	13	2–3.5 cm	12–67 cm	1–2.5 cm	8–20 cm
Toler	2	2 cm	8–12 cm	---	---
Ware	38	2–3 cm (2 cm)*	6–45 cm (10–20 cm)	---	4–74 cm (8–17 cm)
Ewen	16	1–2 cm	8–57 cm (8–23 cm)	1–2 cm	4–35 cm (4–10 cm)
Pilot Knob	5	1.5–2 cm	13–18 cm	---	3–4 cm
Totals/Ranges	86	1–3 cm (2 cm)	6–67 cm (8–23 cm)	1–2.5 cm	3–74 cm (8–20 cm)

*Numbers in parentheses indicate the most frequent size ranges.

The Ware Quarry contained 38 tool marks on boulders and 80 tool marks on millstones. Tool marks were documented on seven boulders (Numbers 1, 2, 5, 13, 15, 18, and 29). Of these marks, 29 were vertical and 10 were horizontal. The vertical tool marks are shallow linear depressions (2 cm wide, except for one 4 cm wide). They range in length from 4 to 35 cm (most 10–20 cm) and are spaced between 4 and 74 cm (most 4–16 cm). The horizontal tool marks are shallow linear depressions that are 2 cm wide. They range in length from 16–45 cm (most 18–30 cm) and are spaced between 9 and 22 cm. Many of these tool marks are parallel in orientation while others are random in placement. Twelve millstones (Numbers 3, 6, 8, 11, 16, 17, 18, 22, 23, 24, 26, and 28) had linear tool marks. The horizontal linear tool marks (n=74) ranged in number from one to fifteen. Two specimens had radial patterns (Millstones 17 and 23). Available size ranges for the horizontal marks were 4 to 80 cm long (most 10–42 cm), 1 to 3 cm wide (most 2 cm), 2 to 3 cm deep, and 4 to 26 cm apart (most 16–17 cm). Vertical tool marks (n=6) occurred in groups of one and four, were 18 cm in length, 2 cm wide, between 12 and 14 cm and 5 to 21 cm apart.

The Ewen Quarry contained 16 tool marks on boulders and 8 tool marks on millstones. Another source of data present on three boulders (Numbers 8, 36, and 39) were the narrow linear tool marks. These tool marks (n=16) were very shallow (1–2 cm) and most ranged from 8 to 32 cm in length. The tool marks were frequently vertically oriented and roughly parallel (4–35 cm apart). Horizontal tool marks on Boulder # 39 ranged in length from 8 to 57 cm long (8, 8, 12, 22, 23, 37, and 57 cm long), 1 to 2 cm wide, and 4 to 10 cm apart (4, 6, 10, and 10 cm apart). Vertical tool marks on Boulder # 36 ranged from 8 to 57 cm long, (10, 13, 15, 15, 15, and 20 cm), were 1 cm wide, and were spaced between 6 and 35 cm apart (6, 8, 10, 22, and 35 cm). Two millstones (Numbers 3 and 4) had linear tool marks. The horizontal linear tool marks (n=3) ranged in number from one to two. Available size ranges for the horizontal marks were 13 to 21 cm long, 1 to 2 cm wide, and 1 cm deep. The five vertical tool marks were observed on one millstone but no measurements were recorded. The seven drill holes were 3–3.5 cm in diameter, between 9 and 13 cm deep, and 22–32 cm apart.

The Pilot Knob Quarry contained 5 tool marks on boulders and 64 tool marks on millstones. Tool marks were observed on two boulders (Numbers 7 and 18). These marks (n=8) were shallow (1 cm deep) linear depressions. They range in length from 10 to 25 cm and are approximately 2 cm wide. Most have a vertical orientation and are parallel. Four millstones (Numbers 2, 8, 9, and 11) had linear tool marks. The horizontal linear tool marks (n=35) ranged in number from four to 16. Available size ranges for the horizontal marks were 5 to 50 cm long (most 10–30 cm), 1 to 4 cm wide, 1 to 3 cm deep, and 4 to 23 cm apart (most

4–12 cm). Vertical tool marks (n=29) occurred in groups of four and 25 and ranged between 8 and 25 cm wide, 2 cm deep, and 8 to 20 cm apart.

Reasons for Millstone Rejection

The millstone maker's hard labor was not always fruitful. At any time during the manufacturing process, a critical mistake could be made or a flaw uncovered inside the rock that ruined the millstone (Table 30). Sometimes a careless blow or a small seam in the rock would cause severe damage to the edges of the millstone. In other instances, the stone would break in an irregular manner when being separated from the parent rock. At other times the millstone would survive until the eye was cut and then the rock would crack or split apart. Regardless of the cause, a rejected millstone meant days of lost labor and much frustration.

Millstones were rejected for several major reasons (Figures 100–101). Often two or three flaws were obvious on the same millstone. At the McGuire Quarry flaws included irregular breaks (n=27), undercutting (n=9), edge damage (n=7), surface depressions (n=7), cracks (n=4), inclusions (n=4), splitting apart (n=1), and unknown causes (n=1). Millstones at the Baker Quarry included irregular breaks (n=19), undercutting (n=2), edge damage (n=2), surface depressions (n=2), cracks (n=4), inclusions (n=1), splitting apart (n=1), and unknown causes (n=1). Problems at the Toler Quarry included irregular breaks (n=10), undercutting (n=3), edge damage (n=5), surface depressions (n=4), cracks (n=5), and unknown causes

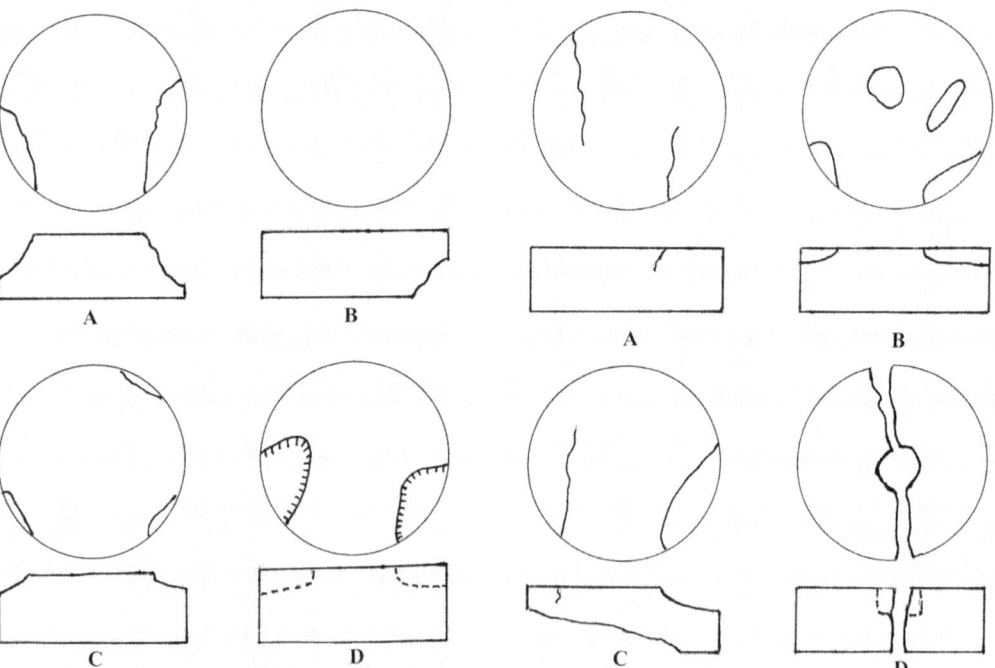

Left: Figure 100. Planview and profile drawings of flaws resulting in millstones being rejected: A, irregular breaks; B, undercutting, C, edge damage, and D, surface depressions. *Right:* Figure 101. Planview and profile drawings of flaws resulting in millstones being rejected: A, cracks; B, inclusions, C, multiple flaws including crack, edge damage, irregular break, and undercutting; and D, splitting apart while cutting eye.

(n=2). At the Ware Quarry flaws included irregular breaks (n=30), undercutting (n=10), edge damage (n=3), surface depressions (n=1), cracks (n=2), inclusions (n=3), and splitting apart (n=2). Millstone problems at the Ewen Quarry included irregular breaks (n=4), undercutting (n=1), edge damage (n=1), surface depressions (n=2), and inclusions (n=1). The Pilot Knob Quarry millstones were rejected because of irregular breaks (n=11), undercutting (n=2), and edge damage (n=1). Overall, the millstones at the six quarries were most commonly rejected because of irregular breaks during shaping and the undercutting of the sides.

Table 30. Reasons that Millstones Were Rejected at the Six Powell County Millstone Quarries

Quarry	Irregular Breaks*	Under-Cutting	Edge Damage	Surface Depressions	Cracks	Unknown	Inclusions
McGuire	27	9	7	7	4	1	4
Baker	19	2	2	2	4	1	1
Toler	10	3	5	4	5	2	0
Ware	30	10	3	1	2	0	3
Ewen	4	1	1	2	0	0	1
Pilot Knob	11	2	1	0	0	0	0
Totals	101	27	19	16	15	4	9

*Many millstone exhibit two or three flaws.

8

Comparisons Between the Powell County Quarries and Other Quarries

The Powell County quarries were undoubtedly a very important source of millstones for grist mills in northeast Kentucky. However, they do not appear to have been of national importance since Kentucky is not listed as a millstone producing state. This omission could be due to several factors. First, the Powell County quarries may have ceased production before the U.S. Geological Survey began collecting statistics in the early 1880s. Second, the Powell County quarries may have been too isolated to economically ship their millstones to distant markets. Third, since good quality conglomerate deposits were available in other eastern U.S. States, the Powell County quarries could not compete with similar millstones locally produced in these states.

This chapter makes some brief comparisons between the Powell County millstone quarries and millstone quarries elsewhere. Unfortunately, few comparisons are possible with other American millstone quarries since they are so poorly known. Some brief comparisons are possible with the British millstone quarries. Traits discussed in this chapter include leveling crosses, reasons for rejecting millstones, quarry excavations, and tools.

Leveling Crosses

In reference to leveling crosses, English millstone expert Owen Ward wrote the author on December 24, 1994: "I should have mentioned that, among other similarities between Pilot Knob and Penallt, you will know that we have some stones with crosses incised across them which Gordon Tucker assumed were cut to assist with the levelling the face of the stone. He was unsure how the depth of each axis of the cross came to be determined. Do you know the explanation, I wonder?"

Gordon Tucker (1977) briefly dealt with leveling crosses. He illustrated millstones with a "level cross" on an unfinished Welch millstone at Penallt, Gwent, and an unfinished millstone at Millstone Edge near Hathersage in England. Tucker (1977:7–8) provided the following comment: "Roughly shaped millstones with crosses cut on their faces have been found both in Wales and the Peak District ... and this suggests that it may have been a common practice for the master millstone mason to cut a level cross to guide a less-experienced man in leveling the main part of the face."

Eight years later in an article entitled "Millstone Making in the Peak District of Derbyshire: The Quarries and Technology," Tucker (1985:51) illustrated a millstone with a leveling cross and made the following comments in the caption:

> Plate 16. Stone cut roughly circular and with crossed grooves cut on the plane of face, at Millstone Edge (at SK 248803). The presumption is that the crossed grooves represent the

beginning of the face-dressing process, with the remainder of the face to be cut by a less-skilled worker. 60 in diameter. Few other specimens like this have been found in the Peak District (only Beely Moor), but similar examples of the process have been found in Wales.

Two millstones with leveling crosses were observed in Pennsylvania and New York. During an April 1998 visit to the Turkey Hill Millstone Quarry in Lancaster County, Pennsylvania, the author observed an abandoned millstone with a partial leveling cross. This unfinished millstone was 105 cm in diameter and 35 cm thick. The leveling cross was T-shaped. One trough was cut across the entire center axis and a second perpendicular trough extended halfway across the other axis. The troughs were 14 to 18 cm wide and 5 cm deep. Another leveling cross was observed on a rejected millstone in Ulster County, New York, during April 1998. This specimen was not measured. These leveling crosses appear to be very similar to the leveling crosses in Kentucky. Two Virginia millstone makers interviewed by Hockensmith and Coy (1999) did not report the use of leveling crosses there.

In summary, leveling crosses were known to have been used in the manufacture of conglomerate millstones in Kentucky, New York, and Pennsylvania. Since they were also used in England and in Wales, it is assumed that the concept of using leveling crosses was probably developed in the United Kingdom. Millstone makers who immigrated to the United States could have easily brought the technology to America.

Reasons for Millstone Rejection

Brush Mountain, Virginia, millstone makers Houston Surface and W. C. Saville shared several reasons for the rejection of millstones (Hockensmith and Coy 1999:36):

Hockensmith: When you are making millstones what are some of the most common problems that would result in you rejecting the stone and what would go wrong?

Saville: Well, if a man held that bull set wrong and pulled too much of your stone off when you are blocking it down, pulled off too much, past your lines, you'd have to throw it away. For you tried to pull off all you could, you know, when you were knocking it off and sometime you'd get too much.

Surface: Yeah, the closer you can [get] to the circle.

Hockensmith: Well, I've noticed some of the ones, the quarries I've looked at, I've documented six millstone quarries in Kentucky and it looks like sometimes when they would be trying to shape the sides, getting it vertical they'd kind of get an irregular break and undercut the stone.

Saville: Yeah, that's what I was talking about, that's where you ruined, when he was holding that bull set because you'd be hitting, maybe a ten or twelve pound sledgehammer and if you'd hold it a little bit too tight, that thing would cut under on you.

Surface: Yeah, it'd go past you round. You see, what we call where your circle is on the side of your stone, it was pretty easy to spoil, wasn't it?

Saville: Yeah.

Hockensmith: Another problem I noticed when they were trying to level the top, sometimes they would get an irregular break or something and part of the surface would be three or four inches lower than the rest of it and they'd just give up because they'd have to take the whole stone down again.

Saville: We never did have no problem with that around here. The biggest problem was when you went to break it down, you know.

Surface: Yeah. And get it past your circle, see, if you go past your circle, see, you ain't got enough stone left to finish....

Hockensmith: And what you all were saying a few minutes ago, one of your problems was when you are putting the eye in, you'd have these hidden fractures in the stone.
Saville: Yeah.
Hockensmith: That wouldn't show up until you were almost done?
Surface: Sure.
Saville: Sometimes when you got finished and were turning it over, it would fall apart.
Surface: See, and all that labor was gone, you see. [laughing]
Hockensmith: Are there any other problems you had in trying, in things that would go wrong when you're making them, that would...?
Saville: Not too much more, no. Mostly, if you got your...
Surface: Sometimes we'd make those things by contract, wouldn't we, Dub?

Millstone makers in Ulster County, New York, reported several problems in manufacturing millstones. The Lawrence brothers mentioned seams in the conglomerate, stones splitting apart, inclusions (geodes), and areas of sand (Hockensmith and Coy 2008a). Irregular breaks were an obvious problem observed in abandoned millstones.

In Lancaster County, Pennsylvania, millstone maker William Nagle told Paul Flory (1951a:77) that "faults" or "seams" in conglomerate caused problems. Undoubtedly, irregular breaks, undercutting, and edge damage were common problems as well.

Quarry Excavations

The rounded pits found at the Powell County quarries were undoubtedly associated with the uncovering of partially buried boulders before they were worked. Flory (1951a:77) provided the following comments based on his discussions with Lancaster County, Pennsylvania, millstone maker William Nagle: "Ofttimes the bowlders would be submerged deeper in the ground than above, necessitating the removal of the earth for a considerable radius about the bowlder before work of 'cutting' or 'trimming' the rock could begin."

The New York and Virginia millstone quarries observed by the author appear to be more oriented to exploiting bedrock. Only the upper portions of the McGuire and Toler quarries would have been comparable. In Ulster County, New York, the conglomerate deposits were exposed on the surface or immediately below the surface. This produced shallow pits where the stone was removed. In Montgomery County, Virginia, workers removed the overburden in the main quarry areas to expose in situ conglomerate. This resulted in larger pits where the stone was worked. Campbell (1925:27) mentioned "many old pits" associated with millstone quarrying at Brush Mountain, Virginia.

Tools

The literature on American millstone making contains few details about the tools used in the manufacturing process. The author had the opportunity to examine actual tools used in the conglomerate millstone industries in New York and Virginia. He also examined some examples found at the Powell County, Kentucky, quarries. Tools from these quarries are briefly discussed in the following paragraphs.

Tools used in quarrying bedrock in Virginia included sledgehammers, drills, wedges and feathers (also called plugs and feathers, wedges and slips, wedges and shims), spoons, and striking hammers (Hockensmith and Coy 1999:51; Hockensmith and Price 1999:84). Specialized hammers included cutting hammers, a bullset, blocking hammers, chipping hammers, bush hammers, and striking hammers. Other small hand tools included points, pitching tools, chisels, and punches. Additional tools included crow bars, wooden calipers, and squares.

New York millstone makers utilized many hand tools to quarry stone and shape millstones. These included bull riggings (wooden holders for drills), sledgehammers, drills, chisels, compasses (made from forked limbs), squares (of different sizes), plugs and feathers, points (of different sizes), crow bars, spoons (for removing rock dust from drill holes), facing hammers, and striking hammers, pitching tools (Hockensmith and Coy 2008a, 2008b).

William Nagle of Lancaster County, Pennsylvania, briefly mentioned some tools in his discussion of millstone making (Flory 1951a:77–79). These tools include "single mouth drills," steel drills, steel wedges, calipers (tree twig with nail on one end and slate held to other end), heavy steel chisels (weighing ca. 40 pounds), sledgehammers, and facing hammers (Flory 1951a:77–79).

The tool kit used by the Powell County millstone makers was probably very similar to the tools used in New York and Virginia. The few tools examined for the Powell County quarries include points, wedges, and feathers. Additional searches at the Powell County quarries would likely result in the discovery of additional tool types.

Limited information was encountered on millstone making and sharpening tools in other countries. Gordon Tucker (1985:50) in his article "Millstone Making in the Peak District of Derbyshire" cited Richard Doncaster concerning millstone making tools. Doncaster obtained a list of early 20th century tools from a man who had been employed in the millstone industry (Tucker 1985:50). These tools include picks, kevels (stone hammers), mauls, punches, wedges, pitchers (iron bars for making holes), hammers, hadges/adzes/axes, plugs and feathers, and reamers (Tucker 1985:50). Hans Werner Jäckle (2002, 2003) discussed tools used in dressing millstones at Saxony, Germany. Also, Mangartz and Pung (2002) discussed the use of wooden wedges in quarrying stone in Germany.

9

Manufacturing Sequences for Millstones

Transforming bedrock or boulders into finished millstones required tremendous skill. Once good quality raw material was obtained (a very important aspect of this work), a series of steps were followed to shape millstones. Available literature suggests that some of these steps varied from location to location, while other practices were nearly universal.

Techniques in Europe

Several excellent accounts are available for British, French, and German millstone manufacture, including some English-language accounts for both France and Germany. Owen Ward (1982a:207) described how the famous French Burr millstones were made at La Ferté-sous-Jouarre, France. He also translated an excellent 1903 account of making French Burr millstones (Ward 1984a:28–29). Several other accounts of making French millstones have been published. These include a description of the techniques at Epernon (Tucker 1982) and another general account (Ward 1982b). Alain Belmont (2006) has recently published a two volume book (in French) on the millstone industry of France. For a general overview of the French millstone industry (in English) and an extensive bibliography of publications on the millstone industry, the reader should see Belmont and Hockensmith (2006). Kenneth Major (1982:196–197) described the manufacture of millstones in the Eifel Region of Germany. For the Eifel area of Germany, the German-language reader should consult books by Fridolin Hörter (1994) as well as Harms and Mangartz (2002). Also in German, a book by Dankmar Leffler (2001) documents the millstone industry at Crawwinkel, Germany. A recent summary by Mangartz (2006) was translated into English.

While we have good summaries of millstone making in France and Germany, the techniques in these countries are quite different from those used in the United States. Thus, the information below will focus on the conglomerate millstones made in Great Britain, where techniques were similar to those used on American soil.

In his 1977 article "Millstones, Quarries, and Millstone-Makers," D. Gordon Tucker (1977:7–8) provided the following description of monolithic conglomerate millstone manufacture: "In some cases it seems the millstone was roughly shaped out of the natural rock before being detached from it; more usually a suitable block of stone was first detached (or found lying by the outcrop) and shaped on the spot. The evidence seems to indicate that even the final shaping was done at the quarry. For this purpose the block might be supported on pieces of stone to raise it above the ground."

A more detailed account of English millstone making was presented in Gordon Tucker's (1985) article "Millstone Making in the Peak District of Derbyshire: The Quarries and the Technology." His discussion of manufacturing methods in central England included five steps inferred from abandoned millstones (Tucker 1985:47): Initially, good quality stone without any obvious flaws was selected and a slab was separated from the parent material. The first stage consisted of leveling the upper face of the slab or block. A small reference hole was placed in the center of the block, a circle was outlined, and the stone was rounded. Stage two consisted of cutting the eye halfway through the stone block. The third stage involved turning the slab over and making this face convex. Stage four consisted of cutting the eye the rest of the way through the slab. Finally, the sides of the millstone were made completely rounded.

Other brief descriptions have been published for millstone manufacturing in the Peak District (Radley 1966:166) and in Anglesey (Tucker 1980:19). In his article "Millstones and Millstone Quarries in Northumberland," George Jobey (1986:60–61) described a method of cutting a round channel around the circumference of the millstone and wedging the millstone from the parent stone. However, the techniques discussed by Jobey (1986) are more similar to methods used at early quarries elsewhere in Europe.

Techniques in the United States

Although many American millstone quarries are mentioned in the geological and historical literature, very little has been published on millstone manufacturing techniques. A brief description was located for the New York quarries and a more detailed account of the methods used at the Cocalico millstone quarries in Pennsylvania was found. Also, the author (Hockensmith 1990c) had an opportunity to interview the last two millstone makers in Virginia. These accounts, when taken together, provide excellent comparative information on the manufacture of conglomerate millstones. The following pages will quote from the New York, Pennsylvania, and Virginia accounts.

New York

Phalen (1908:610) provided the following description of millstone manufacture at the Ulster County, New York, quarries:

> The methods employed in quarrying the rock are simple. The rock is pried or split out, advantage being taken of the joint planes, especially the concentric joints. The tools used are the ordinary hand drill, together with plugs and feathers. Blasting is often resorted to, but the charges of powder are usually light. The rough stones thus obtained are quarry dressed and finished, these operations being performed entirely by hand, the chief tools employed being the bull point and hammer. The operation of drilling the "eye" is performed by centering the stone and then drilling from the center of both faces inward. In many stones the eye is square. To fashion a square eye, a round eye is first drilled out and then squared up. A few of the men engaged in the industry make a modification of the regular millstone for use in the grinding of paint. In this modification the ordinary millstone is cut in halves and an iron casting is placed between the halves, which are then banded together by an iron band.

Much additional information has been collected on the manufacture of Esopus millstones in Ulster County, New York (Hockensmith 2008b, 2008b, editor). This information includes transcribed interviews with Vincent and Wally Lawrence (Hockensmith and Coy 2008a). The Lawrence brothers shared memories of assisting their father with millstone making during their youth. Also, Lewis Waruch, whose mother's family included millstone makers, was interviewed about the millstone quarries near Accord (Hockensmith and Coy 2008b).

Pennsylvania

The late Paul B. Flory, a third generation miller and millstone collector, published one of the best accounts of American millstone making. Mr. Flory accumulated the largest collection of millstones in America, which is now divided between the Smithsonian Institution in Washington, D.C., and Flowerdew Hundred in Prince George County, Virginia (Flory 1951a, 1951b). Being interested in the manufacture of millstones, Mr. Flory (1951a:76) visited the area around the village of Cocalico in Lancaster County, Pennsylvania, where the famous Cocalico Stones were made. Eventually, he was directed to William D. Nagle, who grew up in the mountains close to Mount Airy, the former center of millstone making. Flory (1951a:76) commented that Mr. Nagle "worked for several years in his young manhood as a 'helper' in the actual cutting of millstones from the native mountain bowlders. From him I was able to gather all the facts, fictions and technicalities, as to how millstones were made and where they were found."

Flory's excellent account is quoted below with permission (Flory 1951:77–78):

> Generally, a single millstone or a pair of millstones was cut from a single bowlder. Ofttimes the bowlders would be submerged deeper in the ground than above, necessitating the removal of the earth for a considerable radius about the bowlder before work of "cutting" or "trimming" the rock could begin.
>
> Not every bowlder was suitable for millstones; the stone cutter would carefully examine a rock, looking for "faults" or "seams" before starting work. However, even though all precautions were taken to detect "faults" or "seams" in a bowlder, an occasional "fault" or "seam" would be found after several days' work had been expended on a stone; this was discovered when the stone was "split," as later described. A good quality Cocalico millstone was a very hard pebbly conglomerate, the pebbles varying in size from cherry stones to one inch, or one and one-half inches; the "cherry stone" size being the most preferable.
>
> The first step after selecting a likely looking bowlder was to mark it for splitting to desired thickness, for a lower stone or "bed stone" this was usually eight to twelve inches, and for an upper or "runner stone" sixteen to twenty inches. Using single mouth drills, holes would be driven every eight to twelve inches around the outside circumference of the bowlder and about three inches deep; then using special steel wedges about four inches long, carefully and evenly hammered into each drill hole the bowlder would soon "spring" or separate. If the "split" was a success, i.e. not having uncovered any unseen "fault" in the middle, the stone was ready for the next process, that of marking for diameter desired. The standard stone was four feet in diameter, although for special mills and uses stones varied from two feet to four and one-half feet, and a very few as large as five feet.
>
> Using a tree twig with a nail on one end for the center, and a piece of slate held at the other at the distance desired, a circle was described on the face of the rock. Using heavy steel chisels weighing about forty pounds, the work of "trimming" the stone was begun. One man would hold the chisel while one or two men would strike with sledges. The stone would be "trimmed" down one-half of its thickness around the outside circumference, and then the "eye" of the stone, i.e. the center hole, eight to ten inches in diameter if a "runner," or an eight- to ten-inch square hole if a bed stone, would be cut one-half way through; this operation required special tools and very exact measurements, and was done by the foreman or head cutter.
>
> The surface to be used for grinding was then leveled. This was done entirely by "facing hammers." These facing hammers were made of specially tempered steel, with a face of about two and one-half inches square, having about twenty or twenty-four points. All the irregularities had to be literally pulverized into dust until a perfectly level surface was obtained....
>
> The stone was then turned upside down and the trimming completed on the outside circle and the center hole cut through to exactly meet on the other side. The bottom surface of a "bed" stone or the top surface of a "runner" stone required very little trimming

especially if the split had been fairly straight, so when that was completed the stone was ready to be "iron bound."

Virginia

During May of 1990 the author (Hockensmith 1990c, 1999, editor) interviewed Mr. Huston Surface and Mr. W. C. Saville about their work at the Brush Mountain Millstone Quarry near Blacksburg, Virginia. Both of these gentlemen learned the trade of millstone making from their fathers who were also millstone makers. Even though they worked during the last years of the millstone industry, the techniques and tools they used were essentially the same as those used by their fathers. The following information on manufacturing monolithic conglomerate millstones is based on their descriptions of the techniques. The interviews were initially recorded on video and audio tapes. During 1997, these interviews were transcribed, and an edited version is presented below.

Both large boulders and bedrock were exploited for millstones in Brush Mountain quarries. The larger boulders had to be split into usable sizes (see Hockensmith and Coy 1999: 20–23) while in situ conglomerate was uncovered and blasted free (see Hockensmith and Coy 1999:23–26).

Millstone making can be divided into several steps. These step include laying out the circle, straightening the sides, and cutting the eye. Millstone makers Huston Surface and W. C. Saville provided the following comments which are quoted with permission (Hockensmith and Coy 1999:26–30):

> **Hockensmith:** Okay. Could you kind of describe the first step in quarrying a block stone out? Did you just get a rectangular block to start with?
>
> **Saville:** Yeah, you just shoot your boulder apart, you know, get it laid over. Then you take a little drill, three quarters and to cut your corners off of to get a round. If it was big enough for two, you cut it half in two. And you drill them about three or four inches apart, about four inches deep. Then you take a slip and wedge and put down in there and drive them all equal across there to bust it straight.
>
> **Surface:** I remember on them millstones up there, you all block them, blast them, had an air compressor, didn't you?
>
> **Saville:** Yeah.
>
> **Surface:** You used them little monkey drills to drill them. But you had to do all that by hand along here when you use to make the stones, didn't you? The first ones.

Laying Out the Circle

> **Hockensmith:** How did you all lay out the circle of the millstone?
>
> **Saville:** Just cut a twig off of the mountain with a little fork on it and sharpened that fork and put it in the middle and just went around it with a piece of coal rock or something.
>
> **Surface:** Measure it.
>
> **Saville:** Mark it. There towards the last [we] got to taking that center out of a flashlight battery, the little round thing that goes in it [Surface: "carbon"]. Mark them with [that].
>
> **Surface:** Make a mark, we'd use a piece of flat coal, you know. Stick it agin your stick here, you know, like that and just go right around. It's just as simple as can be. You take a rule and measure it, you know, from your point here where it's sitting out here where the length of your rock [is].
>
> **Saville:** Say if you wanted a four foot rock, you'd get you a stick ... with a fork on it, and

sharpen that fork to go down in the hole. Come out two foot and just make a circle around that.

Hockensmith: So, you put a hole in it to start with in the center?

Saville: Yeah, just a little one. Just enough to hold it.

Hockensmith: How large a hole?

Saville: Oh, just enough to hold that stick in there, maybe a quarter inch deep.

Surface: So, your stick won't shift on you, you know.

Hockensmith: You just drill it with a regular drill or something.

Saville: Yeah, just take a chisel and hammer and hit it.

Surface: Yeah, a little point.

LEVELING THE MILLSTONE

Hockensmith: Well, how about leveling the surfaces of the stone? How did you all keep them perfectly flat when you started?

Saville: [Surface chuckles] Well, you usually took a board and sit it up on it, just site it around and marked it. Get this point of the board and that one over yonder and get your marks where to go to. You take a bull set, they called it. And they'd set there, the bull set and then you'd hit it with a big sledgehammer and knock that stuff up all the way around there to get level around it.

Surface: Where you edge is.

Saville: Then you'd just start drilling that face off.

Surface: Drill that face down level. And you'd go from there, use your cutting hammer and smooth it.

Saville: A man holds that about three quarter, about an inch and three quarter steel and you hit it with a sledgehammer and he'd hold it there and you'd just hit that sledgehammer and just drill that face off.

Hockensmith: What I've seen on some of our stones is what I'd guess I call kind of leveling grooves. They would just cut a groove all the way across the stone and get it perfectly level and then they cut another across it at a right angle. Then I guess they'd just chisel off in between them.

Surface: Yeah, I've seen them do that. Ain't you, Dub?

Saville: All I've ever seen them do is just take a board and get you a straight edge right here, take your hammer and chisel and get a little straight part and just get back and eyeball it around there and put your marks. Then you'd take it off when they had the cutting hammers, what was double faced, sharp on both ends. And you'd take that thing and go around that thing and cut it down smooth. Cut all your drill marks and everything.

Surface: We'd take a flat straight board and take a little damp coal dust and put on there and run across that stone, well, it leaves that there coal dust mark on there on the high places. Take a cutting hammer and cut the high pieces down. And then you do that until you get it where it blacks all the way around. It, it's kindly [chuckles] hard to figure, it's what they called black staffing, didn't it? You put that stuff on that board, straight edge and run across this here flat surface, well, you got it down pretty neat, you know. It's got to be pretty close when you are running two stones together there. You can't let them run together, if you've got a knot on there, it's going to hit, see. You got one bed stone and one runner. And ... it's got to run pretty smooth there.

STRAIGHTENING THE SIDES

Hockensmith: One thing I was always curious about is when you are trying to get the sides of the stone perfectly straight up and down, how do you go about doing that?

Saville: Use a square.

Hockensmith: A square?

Saville: Yeah, just lay your square across the face of your rock, drop it down on it.

Surface: But you've got to have your face square first, don't you?

Saville: Got to have it straight. That's the first thing you do is get that face on it.

Surface: Yeah, and you follow it up all around and get you a perfect circle there.

Hockensmith: How do you keep the stone perfectly round when you are doing that?

Saville: Well, you make your face and then you round it off. Then you take a chisel and go around that thing you call a chisel drafter and cut just about that wide all a way around that stone to get it perfectly straight. And you use that stick that made your mark and you've got that chisel draft all a way around it. Then you start from that and start punching it....

Surface: So, you've got something to work to, you know. And you work from the face to the back.

CUTTING THE EYE

Hockensmith: Could you all describe how you put the eye in, did you drill it out, chisel it out?

Surface: Tell him how you do that, Dub?

Saville: You start at that center and made the size eye that is suppose to go. All right then you took the drill and you just cut, follow that circle right around it. And then they'd come back the next time and straighten that up, make it straight. Then they'd go another trip around it and come back and straighten that up again. And then they'd get out three or four trips around it and then they'd knock the middle out of it.

Surface: Yeah, turns over and do the same thing on the back, you know.

Saville: Try to cut half way through or a little more before you turned it over.

Surface: Now, when your hole is through there in that eye, there's where she'd fall apart on you, wouldn't it? If you wasn't careful. You'd always, you'd first break through from the back, from the front to the back cutting the eye, ah, you, you'd just hit the point of your drill light, you know, when you were breaking through and after that if it didn't fall apart on you when you got the hole through then you can take it....

Saville: You had to be pretty accurate when you turned that thing over to cut that from the other side to make it come in there right. [laughter — Surface] You cut down about halfway and then you turned that rock over and then you went from the back of it after you got the back side like you wanted you go down about halfway in there. When you got down there them holes had to match.

Hockensmith: Why were the central holes so large in some millstones?

Saville: The central hole or eye was made to the specification of the client.

Powell County, Kentucky

A reduction sequence was developed for the Powell County millstone quarries using two major sources of information. First, the incomplete millstones left at the quarries provided actual archaeological examples of the reduction stages. By looking at a wide range of millstones in various stages of completeness, it was possible to determine a rough reductive sequence. Second, the interview with Surface and Saville (Hockensmith 1990c, 1999, editor; Hockensmith and Coy 1999) provided a wealth of information and answered many questions about millstone manufacture. Earlier versions of this reduction sequence have been presented elsewhere (Hockensmith 1993a, 1994a; Hockensmith and Meadows 1996, 1997a).

Selecting the Stone

The procedures for making millstones varied slightly depending on the source of the stone. When utilizing large boulders, the first step was to locate a conglomerate boulder that was either slightly larger than the desired millstone diameter or was large enough to remove slabs from. The raw material had to be well cemented together and be free of obvious flaws. When working with in situ stone, slabs larger than the size of the desired millstone were removed. Millstone makers utilized similar strategies for both quarried slabs and boulders. One difference was that larger boulders contained more excess stone to be removed and those millstones produced from them had to be separated from the parent rock.

Both quarrying and shaping activities frequently required the drilling of holes. Typically, a series of holes were drilled between 18 and 31 cm (7.2 to 12.4 inches) apart in a straight line to detach slabs. A special stone drill with a flared bit was probably used for this purpose. The drill was hit with sledgehammers and rotated after each blow (Figure 102). After the holes were completed, wedges and feathers were inserted in the holes. The millstone maker carefully tapped the wedges with a hammer (Figure 103). When the wedges were sufficiently tight, the stone would split in a straight line between the drilled holes.

Figure 102. Workers using sledgehammers and a stone drill to produce a row of drill holes across the top of a boulder (illustration by Gary Adams, Winchester, Kentucky, reproduced with permission).

9 — *Manufacturing Sequences for Millstones* 145

Figure 103. A worker using wedges and feathers to split a boulder. Note the close-up view of the wedge (left) and the feather (right) inserted into the drill hole (illustration by Gary Adams, Winchester, Kentucky, reproduced with permission).

LEVELING THE MILLSTONE

The second step was to level one side of the millstone (Figure 104). This was done with a leveling staff (a board with one straight side) and special hammers. A powdered substance such as coal dust was applied to the straight side of the staff which was rubbed against the top of the millstone. In this manner, all the high spots would be covered with the powder. The millstone maker would then gently tap the high spots with a special hammer to crush the excess stone. This process would be repeated until all the high spots were removed and the stone was almost perfectly level. Another technique used in Powell County was the cutting of a "leveling cross" on the upper surface of the millstone. This entailed chiseling two wide trough-like grooves into the top of the stone. These grooves were at 90 degree angles to each other and intersected in the center. The leveling cross provided level sighting lines across each axis of the millstone. This gave the millstone maker visual reference points to guide him in removing the excess stone from the four high areas between the arms of the cross.

Figure 104. Worker leveling the upper surface of a millstone with a leveling staff and hammer (illustration by Gary Adams, Winchester, Kentucky, reproduced with permission).

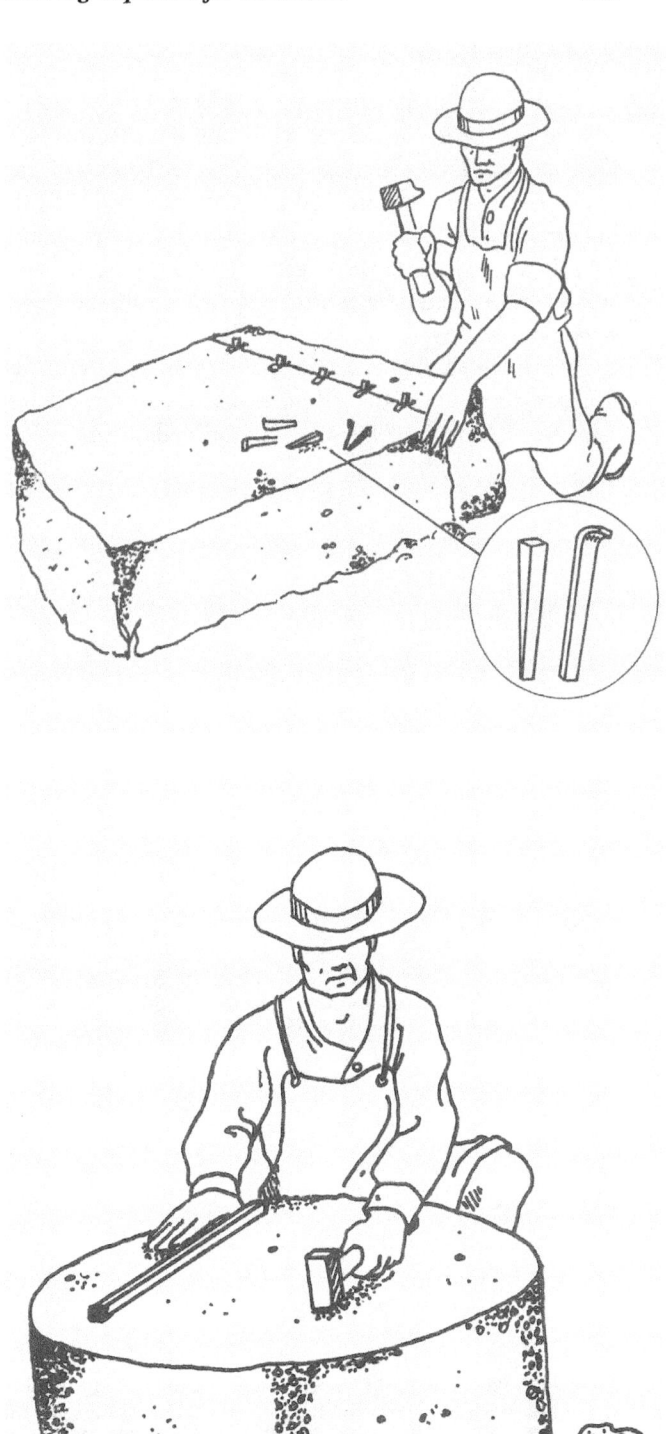

Laying Out the Circle

Once the top of the stone was leveled, a tiny central reference hole was established. Next, a circle the size of the desired millstone was marked with some type of compass. The worker then set out to remove the excess stone to the outer edge of the circle. This could be done in one of two ways. One method probably required the use of a bullset and sledgehammer. The bullset was a large hammer-like tool with a rectangular steel head and a long handle (Figure 105). It was placed on the edge of the stone and hit with the sledgehammer. The amount of stone removed could be controlled by changing the angle and position of the bullset and then striking it with varying degrees of force. The second way to remove excess stone required drilling holes at the corners of slabs or along any unwanted areas of boulders. Wedges and feathers were used to detach the excess stone. After this stage, only a small amount of unnecessary stone remained beyond the outlined circle.

Figure 105. Workers removing the excess stone from a millstone preform. The man on the left is holding a bullset while the man on the right is preparing to strike the bullset with a sledgehammer to remove the angular corner (illustration by Gary Adams, Winchester, Kentucky, reproduced with permission).

Straightening the Sides

The fourth step was to straighten the sides of the millstone (Figure 106). This was done with a variety of special chisels, points, and hammers. The angle of the chisel and amount of force from the hammer controlled the amount of stone removed with each blow. A square

Figure 106. Worker using a hammer and chisel to straighten the sides of a millstone. Note the square on the ground (illustration by Gary Adams, Winchester, Kentucky, reproduced with permission).

was probably used to produce a 90 degree angle between the level top and sides of the millstone. The millstone maker carefully followed the entire circumference of the circle to produce the vertical but curved sides of the millstone. Then the millstone was turned over and the reverse face of the stone was leveled as noted in step two.

CUTTING THE EYE

The final step was to cut the central hole or eye (Figure 107). Available accounts indicate that the eye was outlined and then carefully chiseled halfway through the millstone. The millstone was then turned over and leveled, and the eye was then cut through the remaining stone. Considerable skill was required to ensure that the holes properly met in the middle. The specimens from Powell County suggested that the millstone makers sometimes chiseled the eye in earlier stages. Slight variations may have occurred in the steps followed depending on the millstone maker's experience or preferences.

During the manufacturing process, a misplaced blow or an unseen flaw in the stone could cause a millstone to be rejected. Once the stone broke beyond the circle, the millstone was unsuitable for further reduction. Also, undetected seams, large inclusions, and poor quality stone rendered preforms unusable. Apparently, problems were encountered rather frequently, judging from the number of millstones abandoned at the Powell County quarries.

Figure 107. Worker carefully chiseling the eye (or central hole) into an otherwise complete millstone (illustration by Gary Adams, Winchester, Kentucky, reproduced with permission).

10

Transportation Methods and Routes for Powell County Millstones

An important consideration for millstone manufacturers was transportation. Heavy millstones had to be moved from remote quarries and transported to the purchaser. The rugged terrain surrounding the Powell County millstone quarries made this a formidable task. Since railroads did not serve the area until about 1886, available transportation methods of that day were restricted to sleds, wagons, and flat boats. Sleds pulled by oxen, horses or mules were probably utilized on steep quarry slopes and the primitive roads down the sides of the knobs. It is probable that the more gently sloping finger ridges coming off the knobs were used as routes to reach the valley below. After the millstones were moved to level ground in the valleys, they were most likely transported on heavy wagons.

Road Transport

Most millstones were probably transported on local roads. One historical account mentioned the transportation of millstones in the Powell County area. General Benjamin Logan during his 1796 campaign for governor had the following experience: "On one occasion he was traveling to the town of Winchester to address a gathering. He came upon one of his own wagons which he had sent to the knobs region near Mount Sterling to obtain millstones. The wagon had broken or mired down" (Talbert 1962:284).

Since the Powell County quarries are in the Knobs Region and located roughly between Winchester and Mount Sterling, it is obvious that General Logan was obtaining millstones from one of these quarries. Furthermore, this account indicates that some of the Powell County millstones were being sent to markets in the Bluegrass Region by wagon. It also suggests that some of the purchasers provided their own transportation.

Historical records indicate that a road was built to the millstone quarry at an early date. William Risk's interview noted that a road was cut in the knobs to the millstone quarry about 1793 (John Shane, Draper MSS. 11CC 86). This "Stone Quarry Road" was mentioned in the Clark County Road Order Books for 23 different years between 1801 and 1870 (Enoch and Meadows 2005). For some years the "Stone Quarry Road" was referred to several times and references occurred on many different pages of the Road Order Books. The "Stone Quarry Road" was mentioned in 1801, 1802, 1804, 1805, 1807, 1808, 1809, 1810, 1811, 1813, 1821, 1823, 1826, 1828, 1829, 1834, 1836, 1844, 1855, 1848, 1849, 1865, and 1870 (Enoch and Meadows 2005:5–11, 15–16, 18–19, 21, 23–25, 29–30). The "Stone Quarry Road" was mentioned as a road; references were made to various segments of the road; and it was also mentioned as a landmark where other roads intersected. Also, the "Stone Quarry Road" was mentioned as going to the Mount Sterling Road and the Red River Iron Works Road (Enoch and Meadows 2005).

Land transportation by wagon was undoubtedly restricted to dry seasons since the dirt roads became nearly impassable in the winter and rainy periods. Even after the millstone quarries had ceased production, the roads of Powell County were in poor condition. Moore (1897:138) stated that "there are no turnpikes or McAdam roads in the county, the only roads being what are known as the county dirt roads and they are kept in reasonably good repair." It is also possible that horse drawn sleds could have been used to transport millstones in the winter.

River Transport

It is assumed that millstones going greater distances were transported on the Red and Kentucky rivers. It is not known whether millstones were transported by flat boat on the Red River. However, it is known that the iron industry (active about 1805–1860) utilized the Red River as a transportation route. Verhoeff (1917:158) stated that "in the Kentucky basin early works were located on the Red River near the mouth of Hardwick's Creek in the vicinity of Clay City, Powell County. A dam built at that point rendered the river above navigable for small boats." The millstone quarries were only a few miles north of the Red River and this mode of transportation may have been utilized. In western Kentucky, we know that Shakers at South Union were shipping millstones on the Cumberland River (Arndt 1975:234) and the Barren River (Arndt 1975:712).

A 1799 ad for Red River millstones indicated that they were for sale at Cleveland's Landing on the Kentucky River (Kentucky Gazette 1799b). It is only logical to assume that millstones were transported up and down the Kentucky River on flat boats. Millstones could have been floated down the Red River to the Kentucky River or moved overland to the Kentucky River. Because of the great weight of millstones and the poor condition of early roads, rivers would have been the best routes for transporting millstones over long distances.

Early Spanish documents indicate that millstones were being shipped down the Mississippi River. Apparently, Spanish officials were checking all boats going down the river and making a list of the passengers as well as an inventory of types and quantities of cargo. On April 23, 1790, the Spanish officials at Natchez described the contents of two berchas and one flat boat from Kentucky (Kinnaird 1946:331). One of the berchas owned by William Perkins included six pairs of millstones along with other cargo (Kinnaird 1946:331). It is assumed that these millstones were produced at a Kentucky quarry, possibly the Powell County quarries. They could have easily been floated down the Red River, to the Kentucky, to the Ohio, and finally to the Mississippi River.

Ads published in the *Kentucky Gazette* newspaper document the movement of Kentucky goods by boat to New Orleans. The March 29, 1790, issue of the *Kentucky Gazette* contained the following ad:

> I want to employ a number of hands to conduct my boats to the City of New-Orleans to whom I will give the most liberal encouragement.
>
> JAMES WILKINSON

In the December 8, 1792, issue of the *Kentucky Gazette*, John Moylan published a similar ad looking for men to take his boats down to New Orleans. A number of such ads by different individuals were published in the *Kentucky Gazette* in 1793 and 1794. The reference to millstones being transported in Kentucky boats and the frequency with which boats were being taken to New Orleans suggest a high probability that Kentucky millstones were being shipped to markets to the south.

11

Markets for Powell County Millstones

The full extent of the market for millstones produced at the Powell County quarries is currently unknown. From historical documents, we know that these millstone quarries were frequently referred to, collectively, as the Red River Quarry due to their proximity to the Red River (Hockensmith 2008e). Therefore, it is not surprising that the millstones produced in modern day Powell County would be known as "Red River stones." During a search of the years 1787 to 1820 and selected years afterwards of the *Kentucky Gazette* (Kentucky's first newspaper), several mill ads were discovered that mentioned these millstones. A lawsuit between Fayette County mill owner John Higbee and Absolom Hanks (see Appendix G) indicates that Higbee was trying to obtain Red River millstones as early as December of 1799 (Hockensmith and Meadows 2006). An ad appearing in the September 6, 1803, edition of the *Kentucky Gazette* mentioned that a pair of "Red river stones" was used in the Hickman Mills 11 miles from Lexington. During July 21, 1807, Thomas Hart, Jr., advertised his merchant mill, saw mill and other properties for sale in the *Kentucky Gazette*. Hart's mill on East Hickman, ten miles from Lexington, contained three pairs of millstones including one pair of "Red river" millstones. In the January 10, 1809, issue of the *Kentucky Gazette*, Hart again advertised his mill on East Hickman with the "Red river" millstones. During November 29, 1817, Benjamin Futhey advertised in the *Kentucky Gazette* that his mill on East Hickman which contained a pair of "Red River" millstones and two other types was for sale. Daniel Bradford published a notice in the March 5, 1819, issue of the *Kentucky Gazette* that the Alluvion Mills would be sold. The Alluvion Mills, situated in downtown Lexington, contained two pairs of five foot millstones including "one pair of first quality Red River mill-stones." In the January 2, 1824, edition of the *Kentucky Gazette*, an ad for the old W. H. Tegarden Mill on East Hickman Creek near Tates Creek Road in Fayette County noted that the mill contained two pairs of millstones including one pair of Red River millstones. Finally, the January 2, 1839, issue of the *Observer & Reporter* carried an ad by I. I. McConathy offering for sale "an excellent Pair of RED RIVER MILL STONES, four and a half feet" that had been used in the "Old Lexington Steam Mill."

The above ads suggest that the millstones were used in mills in Lexington and the vicinity prior to 1803. It is suspected that Red River millstones were commonly used in many areas of eastern Kentucky and perhaps in other areas of the commonwealth. While ads for grist mills abound in early Kentucky newspapers, these ads only rarely mention the types of millstones that were used in the mills. Also, since the *Kentucky Gazette* was published in Lexington, it primarily contains ads for the Bluegrass Region. Consequently, it will be difficult to determine the market for the Red River millstones and how far they were distributed. Since only one ad by the quarry operators was found in the Lexington newspapers sampled between 1787 and 1895, it is assumed that the quarry advertised very little. This is surprising since we know that millstone quarries in Franklin and Madison counties advertised in the Frankfort newspapers. Perhaps word of mouth advertising was the preferred method used among the milling community. It is also possible that millstone ads exist in the many years of newspapers that could not be checked due to time constraints or in issues of newspapers that did not survive until the present. Because millstones would last for many years, the market was probably catering to new mills and replacement stones for the older mills.

12

The Competition:
Imported Millstones in Kentucky

In addition to the local Red River millstones, early newspaper ads indicated that the French Burr, Laurel Hill, and Raccoon Burr millstones were also being used in central Kentucky grist mills (Hockensmith 2008c, 2008d; Hockensmith 2008a, editor). The French Burr was regarded by numerous millers as the best millstone in the world. Therefore, it is not surprising that Kentucky millers wanted them. French Burr millstones are listed in the same ads as the Laurel Hill and Red River millstones. The French Burr was undoubtedly used for grinding wheat into flour while the conglomerate Laurel Hill and Red River millstones were more suited to grinding corn. The following paragraphs will discuss ads that mention French Burr, Laurel Hill, and Raccoon Burr millstones.

It is not known when French Burr millstones were first available to Kentucky millers. The earliest ad found to date mentioning French Burr stones was in the May 5, 1792, issue of the *Kentucky Gazette*:

> I HAVE just started a pair of F Burr Stones, for the purpose of grinding Flour; I have good cloths, and a good Miller. Those therefore that will favor me with their custom; if their Wheat is good; may expect good Flour.
>
> I am the Publics
> Humble Servant
> TOLIVER CRAIG.

Four years later, an ad appeared in a supplement of the July 16, 1796, edition of the *Kentucky Gazette*:

> French Bur Mill Stones
>
> THE Subscribers beg leave to inform the public in general and Millers in particular, that they have a quantity of French Bur Mill Stones; which they will dispose of a low rate for Cash. For particulars apply to Henry Strouse, now at Baird & Owen's Store in Lexington or Peter Light at Limestone, we are authorized to sell them.
>
> Jacob & Henry Hoover,
> July 8, 1796

The February 7, 1799, issue of the *Kentucky Gazette* carried another ad for French Burr millstones:

> FOR SALE
> THREE PAIR OF
> French-Bur Mill Stones
> WELL cleaned Merchantable Hemp,
> Tobacco of Superfine flour, will
> be received in payment.
> THOMAS HART

12 — The Competition

> 1st January, 1799
> N. B. A quantity of HEMP is wanting
> for which Nails, Iron or any kind of
> Merchandise, will be given — or Cash,
> payable in six months from the delivery.

Michaux (1805:203) listed French goods coming into Kentucky that included millstones. In the January 23, 1815, issue of the *Kentucky Gazette* newspaper, publisher John Bradford expressed his desire to purchase French Burr millstones:

> CASH
> **For French Burr Mill-Stones.**
>
> WANTED at the Aluvion Mills, Lexington, two or three pair of French Burr Mill Stones of the first quality, and largest size. Any person having such to dispose, will please make known to the subscriber their size and prices as early as possible.
> JOHN BRADFORD.
> Lex. Dec. 12.

The August 5, 1817, edition of the *Paris Western Citizen* included an ad for the sale of Alexander Ogle's grist mill in Bourbon County on Stoner Creek which was equipped with one pair of French Burr millstones. These millstones were 4 feet in diameter and were advertised as "equal in quality to any in America." The entire ad has been transcribed by Larry Meadows (2006:41).

The *Digest of Accounts of Manufacturing Establishments in the United States and of Their Manufactures* (U.S. Secretary of State 1823) mentioned French Burrs in Gallatin County, Kentucky. These French Burrs were mentioned in connection with a flour mill.

Two ads mentioning French Burr millstones were found for 1825 and 1826. A pair of four foot burr millstones were offered for sale by January and Sutherland of Maysville in the February 9, 1825, issue of *The Eagle*. The March 26, 1826, edition of *The Maysville Eagle* carried an ad for the sale of a steam grist mill on Limestone Street in Maysville, Kentucky. The ad mentioned "two first rate pair of 4½ feet French Burr Millstones" among the equipment in the mill.

French Burr millstones were offered for sale in Louisville in the May 15, 1827, newspaper *The Public Advisor*:

> FRENCH BURR MILLSTONES.
>
> One pair, fifty inches diameter, one pair fifty-four inches diameter with spindles & c. complete; expected by first boat from New Orleans. For sale by BUCHANAN & STARKEY

The January 23, 1828, edition of *The Maysville Eagle* carried an ad for the sale of French Burr Millstones. The ad indicated that the Baltimore-based "Morris & Egenton, French Burr Mill-stone Manufacturers," had established a branch office in Cincinnati, Ohio.

H. L. Soper placed the following ad for French Burr millstones in the June 24, 1834, edition of *The Commonwealth* published in Frankfort:

> FRENCH BURR MILLSTONES.
>
> THE subscriber has just received for manufacturing from New York, a very choice lot of FRENCH BURR BLOCKS, which were selected by himself with a great deal of care. Those who may be wanting French Burrs this season, will do well to apply soon — they will be sold at the Cincinnati prices, and warrented of first rate quality, and to be finished in a style equal to any Burrs made in the United States. The subscriber has had many years experience in this business and he assures those who may apply to him for Burrs, that the

work shall come up to what is here promised. All orders in this line directed to me in Madison, Ia. shall be promptly attended to.

H. L. SOPER.

Madison, May 1, 1834 (Ia. Repub.)

The June 1, 1836, edition of the *Lexington Observer & Kentucky Reporter* published the following ad for millstones:

> MILL STONES,
> FOR SALE, a superior pair of 5 foot
> French Burr MILL STONES, and a superior
> pair of 5 foot CORN STONES, with Spindles
> and other machinery for each.
> APPLY TO THE PRINTER

Another source of information on millstones was ads placed for selling mills that listed the equipment contained in the mills. In addition to the term "corn stone" (which usually refers to a local conglomerate millstone), French Burr and Laurel Hill stones are mentioned. Arnow (1984:283) mentioned that a pair of Goose Creek Burrs and a pair of Laurel Hill stones were used in the Croft Mills of General James Winchester on Bledsoe Creek in Russell County when it was sold in 1802. In the April 26, 1803, edition of the *Kentucky Gazette*, an ad for James Morrison's mill near Lexington mentioned "two pair of stones, one of which is Burr of superior quality; the other Laurel-Hill, excellent for country work, or grinding corn." An ad by D. M. Vicar appeared in the September 6, 1803, *Kentucky Gazette* which offered the Hickman Mills, seven miles from Lexington, for rent. The Hickman Mills had "one pair of five feet burrs, and one pair of Red river stones." The September 20, 1803, edition of the *Kentucky Gazette* carried an ad for John Rogers' grist mill on Davy's fork of Elkhorn in Fayette County. Rogers' mill contained "two pair of stones, one of which is French Burr." The following year, John M. Call advised the public in the January 31, 1804, issue of the *Kentucky Gazette* that his mill in Fayette was in complete order and has "a first rate pair of French Burrs." In the March 13, 1804, edition of the *Kentucky Gazette*, John Rogers again offered to sell his mill on Davy's fork of Elkhorn in Fayette County which contained one pair of French Burrs and an unspecified set of stones. William Watson of Clark County offered his Tub Mill on Boon's Creek for sale in the September 3, 1805, issue of the *Kentucky Gazette*. Watson's mill contained "two pair of Laurel Hill stones." Elisha I. Winter, Jr., advertises the sale of his grist mill at the mouth of Tates Creek, 15 miles from Lexington on the road to the Madison County courthouse, in the November 21, 1805 *Kentucky Gazette*. Winter noted that the mill was equipped with "a pair of French Burrs and a pair of Laurel Hill stones." On July 21, 1807, Thomas Hart, Jr., advertised his merchant mill, saw mill and other properties for sale in the *Kentucky Gazette*. Hart's mill on East Hickman, ten miles from Lexington, contained three pairs of millstones including "one pair Burrs, one pair Laurel Hill, and one pair of Red river." John M. Call published an ad in the April 19, 1808, edition of the *Kentucky Gazette* advising the public that his Oak Ridge Mills in Fayette County were for sale. In his description of the equipment, Mr. Call listed "a pair of French burrs four and a half feet, of superior quality (waranted)" and "a pair of country stones." In the January 10, 1809, issue of the *Kentucky Gazette*, Thomas Hart, Jr., again advertised his mill on East Hickman with the same listing of millstones.

During November 29, 1817, Benjamin Futhey advertised in the *Kentucky Gazette* the sale of his mill on East Hickman ten miles from Lexington. Futhey's mill contained three pairs of millstones including "one pair French Burrs, one pair Laurel Hill and one pair of Red River."

12 — The Competition

The August 5, 1817, issue of the *Paris Western Citizen* included an ad for the sale of Alexander Ogle's grist mill in Bourbon County which had "two good pairs of Laurel Hill stones" and one pair of French Burr millstones. In the March 5, 1819, issue of the *Kentucky Gazette* Daniel Bradford published a notice that the Alluvion Mills would be sold. The Alluvion Mills, in Lexington, had the power to "drive two pair of five feet millstones, one pair of superior French Burr, and one pair of first quality Red River mill-stones, five feet diameter." An ad for the sale of the old W. H. Tegarden Mill on East Hickman Creek near Tates Creek Road in Fayette County appeared in the January 2, 1824, edition of the *Kentucky Gazette*. The ad noted that the Tegarden Mill contained one pair of French Burr millstones and one pair of Red River millstones. Francis F. Johnson published a notice in the July 27, 1836, issue of the *Lexington Observer & Kentucky Reporter* of a public sale of his grist mill on Four Mile Creek in Clarke County, seven miles south of Winchester. Johnson's mill had two pairs of millstones including "1 pair of French Burs, and 1 pair Corn Stones." Finally, E. and H. Stedman published an ad in the March 11, 1845, issue of *Frankfort Commonwealth* offering millstones for sale. The Stedman brothers advised that they had one pair of French Burr millstones and one pair of Laurel Hill millstones for sale at their mill in Frankfort. In 1848, Herbert & Wright, Mill Furnishers in Louisville, were offering Laurel Hill millstones for sale (Collins 1848:362). Again, in 1850, Herbert & Wright were selling Laurel Hill millstones in Louisville (Jegli 1850:85).

The suppliers of French Burr millstones to Kentucky mills probably changed through time. Initially, these millstones were brought into the commonwealth by individuals. Our first ad dates to 1792 but French Burrs may have been installed in some Kentucky mills a few years earlier. An account of James Flint's (1904:240) travels between 1818 and 1820 indicate that Cincinnati had "one burr millstone factory" by that time. Webb (1933:33) stated that "many French buhrs reached Kentucky by way of the Ohio River, coming by boat from New Orleans." Verhoeff (1917:66) also stated that millstones from France were shipped to Kentucky from New Orleans on keel boats.

In 1848, Herbert & Wright, Mill Furnishers in Louisville, were offering all sizes of French Buhr millstones for sale "made from best quality Buhr Blocks" (Collins 1848:362). Again, in 1850, Herbert & Wright were selling French Buhr millstones in Louisville (Jegli 1850:85).

During 1851, additional information was provided about French Burr millstones being manufactured in Cincinnati. Cist (1851:182) provided the following information:

> *Burr Millstone makers.* Four factories.—Nineteen hands; value of product, twenty-four thousand dollars; raw material, 65 per cent.
>
> James Bradford & Co., 65 Walnut street, manufacture yearly, seventy-five pairs burr millstones.
>
> The burrs, of which the millstones are composed, are imported from France, in cubes of about twelve inches average. We have the same material in our own west, but it is not hard enough for service. The burrs are cemented with plaster of Paris, which is received from Nova Scotia and the Lake Erie region; and each stone is secured with four bands of iron, which being put on hot, as they shrink in cooling, serve to confine the whole under any amount of strain to which it may be exposed.

Ten years later, in 1861, T. Bradford and Company of Cincinnati were selling French Burr millstones and many other supplies for mills (Williams and Company 1861:480). They published the following ad in the 1861 Covington, Kentucky, city directory:

> **FRENCH BURR MILL STONE MANUFACTORY.** Genuine Dutch Anker Brand Bolting Cloths, of all numbers, Mill Castings, Mill Spindles, Hoisting Screws, Damsel Irons,

Tempering Screws, Screen Wire, Smut Machines, Plaster Paris; always on hand and for sale.

T. BRADFORD & CO.

Office, 59 Walnut, Factory 135 West 2d St. bet. Race and Elm.

The value of the French Burr millstones can be estimated from a few ads. The January 17, 1852, issue of *Scientific American* carried an ad for a run of 4½ feet diameter French Burr Stones to be put on a ship and sent to the purchaser for $145 (Scientific American 1852:143). The April 11, 1863, issue of *Scientific American* ran an ad for four pairs of French Burr stones being sold for $100 per pair in the hoops (Scientific American 1863:238).

In addition to the French Burr and Laurel Hill millstones, English millstones may have been used in Kentucky. Webb (1933:33) states that English conglomerate millstones were brought into Kentucky by early settlers. These were similar to American stones but were more carefully dressed on the non-grinding face. Apparently, English millstones were brought to America as ballast on sailing ships (Webb 1933:32). After 1810, English millstones did not come into Kentucky (Webb 1933:33). These English millstones were transported from Virginia by oxen drawn cart along the Wilderness Road (Webb 1933:32). Due to the great difficulty and cost of transporting the stones, it is unlikely that many English millstones were brought to Kentucky.

Raccoon Burr millstones have been mentioned in early newspaper ads. A September 14, 1825, ad appearing in *The Eagle*, published in Maysville, discussed the sale of the newly erected water grist mill and other property which part of the estate of David Johnston. The mill contained "two pair of excellent Raccoon Burrs." Nine years later, *The Maysville Eagle* ran the following millstone ad on November 20, 1834:

MILL STONES.

RACOON BUR MILL STONES, manufactured by David Richmond, of various sizes, warranted. Can be had at short notice, by application to

JANUARY & HUSTON, Agents

Maysville, May 3, 1834

In 1850, Herbert & Wright were selling Raccoon Buhr millstones in Louisville (Jegli 1850:85).

The Raccoon Burr millstones were manufactured in Ohio (Hockensmith 2003c, 2007). A millstone quarry was established about 1805 by a man named Musselman in Vinton County (Garber 1970:78). These millstones became known as "Raccoon buhrs" since the quarry was near Raccoon Creek (Garber 1970:78). Blocks of stone were quarried and fitted together to form composite millstones (Cinadr and Brose 1978:59). During 1853, the Raccoon quarry was owned by Walden & Ward (Garber 1970:79). By 1860 the Raccoon Creek quarry had apparently ceased production "either due to exhaustion of the supply of quality material, or because of the difficulty and expense involved in production" (Garber 1970:79).

As this chapter has indicated, central Kentucky millers had several choices of millstones available to them. The Red River stones could not compete with the famous French Burr stones. However, the Red River stones were probably very comparable in quality to the Laurel Hill stones, which were also made from conglomerate. Likewise, the English millstones were made of conglomerate. The Raccoon Creek Burrs were made from flint and were thus more similar to the French Burr stones. In light of the competition, the market for the Red River millstones was undoubtedly restricted to grinding corn. The French Burr or Raccoon Creek Burrs would have been used for grinding wheat.

13

Kentucky Millstone Values

Little information is available about the prices of millstones quarried in Kentucky. Many of the ads found do not mention the prices but only sizes of millstones available. Excellent ads were found for flint millstone quarries in Franklin County during 1831. Two ads, published in 1799 and 1818, were discovered for Red River millstones but only one of the ads included the prices. It is interesting to note that cash was not the only way to pay for millstones. Horses were accepted at Cleveland's Landing while whisky was accepted in Rockcastle County.

During 1821, two millstone quarries in Franklin County, Kentucky, were competing for customers. The firm of Miller, Railsback & Miller were selling millstones by diameter. Their competitor, Jeremiah Buckley, priced his millstone by diameter as well. The firm of Miller, Railsback & Miller advertised their flint millstones from a quarry in Franklin County in the August 9, 1821, edition of *The Argus of Western America*:

5 feet	$ 150.
4 feet	"100.
3 feet	"50.

And all sizes accordingly.

Jeremiah Buckley offered greater size options for his flint millstones as advertised in the November 8, 1821, edition of *The Argus of Western America*:

The prices of my mill stones are as follows:

For Five Feet	$180
Four feet six inches	150
Four feet	125
Three feet nine inches	100
Three feet six inches	85
Three feet three inches	75
Three feet	60
Two feet nine inches	50
Two feet six inches	40

The June 13, 1799, edition of the *Kentucky Gazette* contained an ad for the sale of five pairs of Red River millstones at Cleveland's Landing:

RED-RIVER MILL-STONES

FOR SALE, at Cleveland's landing, five pair of Red-River Mill-Stones of the best quality from that quarry, of the following size, viz.— 4 feet,— 3 feet 10 inches,— 3 feet 8 inches,— 3 feet 6 inches,— 3 feet, in diameter.— Cash or good horses will be taken in payment.

Likewise, Lease of 200 acres of Land, lying on the West fork of Howard's creek, two miles from the stone Meeting house, for three years, (including the present,) together with the growing crop, consisting of 30 acres of corn, about 6 of tobacco, &c. Also, stills well fixed on distilling, with all conveniences appurtenant thereto. Apply to the subscriber, on the premises.

William Gordon

June 10, 1799

A second ad for Red River millstones was published in the April 1, 1818, issue of the *Kentucky Reporter*:

MILL STONES.
(OF A SUPERIOR QUALITY,)

WILL in future be made at the RED RIVER QUARRY and sold by SPENCER ADAMS & JAMES DANIEL. If any of our Mill Stones should not prove good, we bind ourselves to furnish more at the quarry until they do prove good. The prices of Stones at the quarry is as follows, to witz — For five feet stones 200 dollars per pair; for four feet 150 dollars; for three feet 80 dollars; all other sizes in the same proportion as above. The prices of a runner will be two thirds the price of a pair. All persons wishing to purchase will apply to the undersigned in Winchester Ky either personally or by letter, which will be thankfully received and duly attended to.

JAS. DANIEL

March 26, 1818

On April 18, 1832, Brutus Clay paid William Rodes $40 for millstones (Clay 1994). It is not known what type of millstones these were.

The limited information available suggests that millstones were very expensive. Undoubtedly, this was due in part to the great skill required to make them and the difficulty in transporting them on primitive roads. The flint millstones sold for $40 to $180 each. The Red River millstones were valued at $80 to $200 per pair.

14

Conclusions

The geological literature and the extensive quarry remains indicate that the quarries in Powell County played an important role in supplying millstones to markets in central and eastern Kentucky. The archaeological study of these quarries has provided new insight into this poorly known industry. While many questions remain unanswered, important details about millstone manufacture have come to light. We now know how these millstones were manufactured, how the stone was uncovered and worked, and some of the sizes of millstones produced.

Archival research has also identified some of the individuals associated with the Kentucky millstone industry: millstone makers Spencer Adams, James Daniel, Absolom Hanks, Martin Johnson, Peter DeWitt, Sr., Cornelius Spry, Peter Treadway, and Moses Treadway, and quarry owners Spencer Adams, James Daniel, Peter DeWitt, Sr., Anderson Pigg, Cornelius Summers, and Moses Treadway. Undoubtedly, many other men worked at these quarries during the several decades they were operating.

The lawsuits that mentioned the Red River millstone quarries offer insights into their size, quality, payment, and delivery. It appears that the quarry owner would take orders for specific diameters and thicknesses of millstones. The 1804 lawsuit mentioned millstones four feet in diameter with the runner stone being 20 inches thick and the bedstone being 12 inches thick. In 1810, millstones were mentioned as being three feet in diameter, with both the runner stone and bedstone being 17 inches thick. A third lawsuit mentioned millstones as being four feet in diameter and the usual thickness. There was also an emphasis on the quality of the millstones, promising stones of "good quality," "best quality," made in a "neat and workmanlike manner." The millstones were guaranteed with the understanding that replacement stones would be cut if the first ones did not prove good. The millstone maker was to ensure delivery to the client. Payment for the millstones was made either in currency or in whiskey. Also, the Red River millstones were mentioned in connection with grinding corn.

Other details about the Red River millstone quarries were mentioned in the lawsuits. The usual season for making millstones was between April 1 and the end of November. Martin Johnson furnished his own tools at the millstone quarry; it may have been common for millstone makers to bring their own tool kits. The labor agreements seem to have varied. Some employees were paid on the basis of the length of time that they worked and also received boarding from the quarry owner. Martin Johnson, on the other hand, lived in his own house and was to be paid one-fourth of the quarry's income after expenses. During 1819, Spencer Adams' quarrying operation took in $1,500, of which $1,200 remained after expenses. The lawsuit of Johnson versus Adams indicated that Spencer Adams maintained books on the quarry.

Finally, the lawsuits indicate that the quarries were in Montgomery County in 1804 but that the line changed before 1810, which placed the quarries in Clark County. The quarries became part of Powell County in 1852 when that county was established.

Several factors were important for the success of a millstone quarry. First, suitable stone had to be available in sufficient quantities to be commercially exploited. Second, skilled workers were essential, and they had to be equipped with the necessary tools. Third, a knowledgeable blacksmith was needed to sharpen and temper the many specialized tools. Fourth, there had to be a market for the millstones produced. Fifth, cheap transportation was very important. The Powell County millstone quarries appear to have had all the important human and natural resources that were necessary.

Unfortunately, the loss of important deeds for the quarry properties, during a courthouse fire in the Civil War era, left gaps in the archival record. Many key questions remain to be answered. There are questions about business operations: Were the quarries operated by different individuals at the same time or by successive millstone makers through time? Were the quarries owned or leased by the millstone makers? Were the millstone makers full time in the trade or did they just make millstones only when they had orders? A second area of questioning deals with the age of the quarries. Were the quarries all contemporaneous or do they represent a series of quarries operated through time as conglomerate deposits were depleted? We know from available records that millstones were being quarried as early as the 1790s and until some time in the 1870s. Were the boulders on the slopes initially exploited for millstones and the in situ deposits uncovered later, higher on the knob (McGuire and Toler quarries)? Hopefully, future research will find the answers to these and other questions.

It is suspected that the six quarries represent only a sample of the existing quarries in northwest Powell County. Since the knobs are very rugged and heavily vegetated, a great deal of time and effort would be required to explore the numerous little valleys dissecting this area. Undoubtedly, additional millstone quarries will come to light as this portion of Powell County is intensively surveyed. The study of new quarries will enhance our understanding of the millstone industry in Powell County.

Undoubtedly, a re-examination of the Powell County millstone quarries in light of current information could yield new insights. When this research was initiated in the spring of 1987, little was known about the American millstone industry and nothing was known about the Kentucky millstone industry. There were no archaeological studies to consult to find out what types of remains researchers should look for in the field. As this work progressed, the author learned more about the millstone industry and was able to see additional evidence of quarrying during each field season. An unfortunate reality of fieldwork is that archaeologists usually tend to see only what they are looking for. A finished millstone is easily recognizable but most people would not see tool marks on rough conglomerate boulders. Thus, only archaeologists that are well acquainted with the conglomerate millstone industry can ably re-evaluate the quarry remains.

It is hoped that future research can build upon the information reported in this book. Given the dense vegetation and thick leaf cover, more remains are probably hidden at each quarry. In fact, it would be very difficult to identify all quarrying remains because of the vegetation obscuring it, stones being partially exposed, and some stones being buried by silt along streams. It would also be useful to have detailed contour maps showing precisely where the remains are located at each quarry. The quarry remains could be reassessed and additional details could be recorded. At some point in the future, researchers may wish to turn the millstones over and document the other sides. Interviews conducted in New York and Virginia suggest that several blacksmith shops were connected with the quarries. (Stoneworking tools required constant maintenance to cut conglomerate.) Intensive survey of the quarries could locate the remains of these blacksmith shops. Stoneworking tools could probably be located by systematically covering the quarries with metal detectors.

Additional archaeological investigations at the quarries may yield new data. It would be interesting to excavate trenches on the benches where the bedrock was being quarried at the tops of the knobs. Potentially, excavations could expose the working surfaces with extraction depressions and drill holes. Tools and other artifacts may be recovered as well. This would increase our understanding of the quarrying activities at the outcrops of conglomerate. Small pits could be better understood by conducting excavations in them. Boulders are still visible in some of these pits. It is assumed that the pits represent holes dug to expose and work around large boulders. Thus, the excavation of pits could shed light on the activities occurring in these features. The linear trenches, located near the knob crest, should also be sampled by excavation. It is assumed that these depressions were produced by the millstone makers exposing in situ layers of conglomerate. Finally, blacksmith shops and temporary shelters for workers may be located that could be sampled by excavations.

This study has shown the research potential that quarry sites have for addressing questions about the manufacture of millstones. Since few historical accounts exist for this industry, our understanding of millstone quarrying must be largely derived from future archaeological studies. Consequently, I would like to encourage other archaeologists to seek out and study millstone quarries in their respective states. Only when the archaeological community realizes the research potential of millstone quarries will we begin to understand this industry that was once so vital to America.

Glossary

There are several specialized terms and names that are connected with the millstone industry. Some of these terms refer to the tools used in the manufacture and sharpening of millstones. Most of these terms are not in common usage in every day language; many relate to names given to specific brand names of millstones, and others are general millstone terms, quarry terms, and types of tools.

Bedstone The lower millstone that remains stationary during the grinding process.

Blocking Hammer Stone working hammer used to removing large pieces of stone during the initial shaping of millstones. These hammers had rectangular heads with square striking surfaces and came in 6 pound and 16 pound sizes.

Boulder Large pieces of conglomerate that are usually scattered across the landscape and not part of an outcrop. In Powell County, the term has been used to designate large pieces of conglomerate found at the millstone quarries.

Brush Mountain Millstones Formerly well known white conglomerate millstones made on Brush Mountain in Montgomery County, Virginia, near Blacksburg.

Bullrigging A segment of a small hickory sapling that has been notched in the center. A drill or other tool was held where the sapling was folded over at the notches. This wooden tool holder served as an extension of the arm to allow the worker clearance from sledgehammers hitting the drill.

Bullset A bullset was a hammer-like tool with a long handle that was struck with a sledgehammer to remove large chunks of stone during the initial shaping of millstones.

Burrs A term referring to millstones or grinding stones.

Burrstone A stone with suitable characteristic for being used as a burr millstone. This often refers to a porous flint type stone.

Bush Hammer A small stone working hammer with square striking surfaces containing rows of pointed teeth. These hammers were used to pulverize the high spots on the grinding faces of millstones.

Caliper A tool used to measure the diameter of a millstone. Simple calipers could be made from a forked branch cut to the desired length. The short side of the fork was placed in a shallow central hole and a piece of coal was held to the opposite end to outline a circle as the caliper was rotated.

Chasers Vertical running millstones attached to a shaft that that turned in a circle to crush ores and other materials. One stone chases or follows the other stone around the circle.

Chipping Hammer A stone working hammer that had wedge-shaped striking surfaces on both ends.

Chisel An iron bar with a sharp flat cutting blade on one end. It is hit with a hammer on the other end.

Cocalico Millstones Conglomerate millstones once made in Lancaster County, Pennsylvania, near the Cocalico Township.

Composite Millstones Millstones made from several carefully shaped blocks of stone that were cemented together. Iron bands were placed around these millstones to help hold them together.

Conglomerate A type of sandstone containing rounded quartz pebbles. The color of the sandstone and size of pebbles can vary greatly. These geological deposits are usually associated with ancient stream beds where the sand and pebbles become cemented together to form rock.

Cross Grain Grain refers to the bedding plane in conglomerate. When the stone was broken across this grain, rather than with the grain, the resulting break will be irregular.

Cutting Hammer A special stone hammer resembling a double bitted axe that was used in leveling the face of a millstone. They were made in different sizes to accommodate different diameters of millstones.

Derrick A device for lifting and moving stones at a quarry. They were usually composed of a large vertical timber secured with guide wire connected to the top. A hinged timber boom was attached to its base that could be raised and lowered with a crank. The derrick also rotated to facilitate the movement of stone.

Drill A steel rod with a flared cutting bit for drilling holes in conglomerate. The drill was hit with sledgehammers and rotated between blows. Longer drills with smaller bits were used as the holes became deeper.

Drill Holes This term has been used in Powell County to refer to either complete holes or the profiles of drill holes remaining after conglomerate was split.

Edge Runners Millstones designed to operate vertically on their edges or sides. They range from single stones working alone to two stones, attached to the same axle, that follow one another.

Esopus Millstones Famous white conglomerate millstones made at several communities in Ulster County, New York.

Eye The central hole in a millstone where the grain is fed in and where the shaft is attached to turn the runner stone. Eyes in runner stones are usually round while eyes in bedstones are usually square. Notches were often added on either side of the eyes on runner stones to facilitate their attachment to the power source.

Face Grinders Millstones designed to work in pairs and grind in a horizontal position.

Face of Millstones The leveled grinding surface of a millstone on which the furrows are cut.

Flint Ridge Millstones Early monolithic and later composite millstones made from flint in Licking and Muskingum counties, Ohio.

French Burr Millstones Millstones quarried at La Ferté-sous-Jouarre east of Paris and other nearby areas of France. The early French millstones were monolithic while the later millstones were composite.

Furrows Shallow grooves cut into the grinding surfaces of millstones to facilitate the grinding of grains as the upper stone rotated. Several patterns of furrows were used by millers.

Goose Creek Burrs Millstones possibly produced in western Kentucky in the early 19th century. The name Goose Creek may refer to the former name of a stream near the quarry.

Laurel Hill Millstones Conglomerate millstones once quarried on Chestnut Ridge in the Allegheny Mountains of Fayette County, Pennsylvania.

Leveling Crosses Two trough-like depressions at right angles on the top (face) of an unfinished millstone that intersect in the center of the stone. These crosses provided level surfaces across both axes of a millstone. They also provided sighting lines for removing the pie-shaped high areas in between.

Meuliers The French word for millstones.

Millstone Cutters Individuals that cut or shaped millstones at a quarry.

Glossary

Millstone Factory A factory, usually in an urban area, where composite millstones (usually imported French Burr pieces) were shaped, assembled, and sold.

Millstone Grit An early geological term that refers to conglomerates since these stones have long been connected with millstones.

Millstone Pick A tool for cutting furrows in millstones. This hammer-like tool had a horizontal head resembling a chisel that was sharp on both ends.

Millstones Disk-shaped stones made for the grinding and/or crushing of grains, ores and other materials. They were commonly made from conglomerate, granite, and flint in the United States. Millstones were usually used in pairs.

Mineral Resources of the United States An early series of annual reports on minerals in the United States published by the U.S. Bureau of Mines.

Minerals Yearbook Annual series of reports by the U.S. Bureau of Mines on U.S. mineral industries. It was a successor to the earlier Mineral Resources of the United States.

Monolithic Millstones Millstones made from a single piece of stone.

Moore County Grit Millstones Blue granite millstones with white flint inclusions once quarried in Moore County, North Carolina, near the community of Parkewood.

Outcrops Exposed areas of in situ bedrock that can be exploited after being uncovered by excavation.

Paint Staff A straight edged board that was coated with pigment to rub across the grinding face of a millstone. The high spots were coated with pigment and were worked down to level the grinding surface.

Peak Millstones English monolithic millstones quarried from millstone grit in the Peak District of Derbyshire.

Pitching Tool A chisel-like tool with a very blunt cutting edge. It is hit with a hammer to remove large pieces of stone while shaping a millstone.

Pits In Powell County, the term pit is used to refer to shallow oval depressions that were associated with excavations to expose the bases of large boulders.

Plugs and Feathers Plugs are metal wedges that fit between two L-shaped feathers placed on either side of drill holes. As the wedges are gradually taped, the pressure is increased on the sides of the holes causing the stone to split in a line along the drill holes.

Point A short iron rod that is tapered on one end (on four sides) to form a pointed cutting edge. This tool was hit with a hammer to remove excess stone during the shaping of a millstone.

Proof Staff A metal staff used to check wooden paint staffs for straightness.

Quarry An area where bedrock or boulders were removed by splitting with hand tools or by blasting. If substantial amounts of stone was removed, a depression was produced in the landscape.

Querns Small millstones designed to be turned by hand power.

Raccoon Creek Buhr Millstones Composite flint millstones made in Athens, Licking, Muskingum, and Vinton counties, Ohio. Their name is derived from Raccoon Creek that flowed through the area.

Red River Millstones Light brown conglomerate millstones quarried in present day Powell County, Kentucky, near Pilot Knob or Rotten Point. They were named for the Red River, which drains this area.

Roller Mills Mills that used grooved steel rollers to grind grain. This new technology lead to the demise of the millstone industry.

Runner Stone An upper millstone that rotates during the grinding process. A shaft attached the millstone was connected to the power source to turn it.

Sandstone A sedimentary rock primarily formed by grains of sand that are cemented together.

Shaping Debris Pieces of conglomerate chipped or broken from millstone blanks and boulders during the shaping process. These discarded rocks were either left where they fell or piled out of the way.

Sharpening Millstones Recutting furrows in millstones that have become worn or dulled by frequent grinding.

Sledgehammer A large heavy hammer (usually eight pounds or more) used to strike a rock or another tool. These hammers can be blunt on both ends or blunt on one end and wedge-shaped on the other end.

Stone Cutter An individual that cuts and works stone at a quarry. The millstone makers were usually known as stone cutters.

Striking Hammers Hammers with blunt faces that were used to hit chisels, points, and other tools.

Tool Marks Visible marks in conglomerate resulting from tool use. These can be drill holes, narrow linear lines or other types of marks.

Turkey Hill Millstones Monolithic conglomerate millstones once made on Turkey Hill in Lancaster County, Pennsylvania.

U.S. Bureau of Mines and Minerals The former federal agency charged with the collection and reporting of information on U.S. mines and mineral resources. The U.S. Geological Survey in Reston, Virginia, currently maintains information on mineral related industries.

Appendices

A. Form for Documenting Millstones

MILLSTONE RECORDING FORM

Location and General Data

Field Specimen Number: _____ Current Location:_____
_____

County: _____ State: _____

Setting of Millstone: _____ streambed, _____ quarry, _____ grist mill, _____ base of hill, _____ top of hill, _____ slope of hill, _____ yard of residence, _____ yard of business, _____ other (_____).

Original location (if moved after original use):_____
_____

Owner: _____ Owner's Address:_____
_____

Archaeological Site Number: _____

U.S.G.S. 7.5 Minute Topographic Quadrangle: _____

Description

Raw Material: ___ conglomerate, ___ sandstone, ____ chert, _____ French Burr stone, _____ limestone, ___ other (_____)

Description of raw material including texture, inclusions, color, size and color of pebbles, etc.
_____
_____

Measurements; Diameter: ___ cm (___ inches), Thickness: ___ cm (___ inches)
    Diameter of round hole (eye): ___ cm (___ inches), Size of rectangular hole (eye):
    ___ cm (___ inches) by ___ cm (___ inches)

Distance from edge of stone of edge of central hole (eye): ___ cm (___ inches)

Other shapes of central holes (+, etc.) _____

Type of construction: ___ one piece of stone, ___ composite stone

Composite Millstone Data

Describe the shapes, sizes, and number of individual stones (shaped segments) used in making the millstones: _____
_____
_____

Description of iron band(s) around the millstone:

      Width: ___ cm (___ inches), thickness: ___ cm (___ inches). If more than one iron band is present, describe: _____
_____

Are the composite stones cemented together? If yes, describe: _____
_____

Is the back of the stone filled with plaster of Paris:_____

Description of Millstone Dressing

Has the millstone been dressed for grinding? _____ If yes, describe (sickle, two-furrow, four quarter, etc.) and attach sketch:_____

How many grooves (furrows) are there? ____; How wide are the grooves? ____ cm to ____ cm; How deep are the grooves? ____ mm; How for apart are the grooves?: in the center (next to eye): ____ cm to ____ cm; at the outer edge: ____ cm to ____ cm; and midway between the center and the edge: ____ cm to ____ cm.

Describe the grinding surface: _____ flat, _____ concave, or _____ convex.

Does the stone exhibit wear from use? If yes, describe: _____
_____

Does the millstone retain portions of the gears (rynd, mace head, damsel, spindle, etc.)? If yes, describe:
_____
_____

Unfinished Millstones

Describe the degree of completeness: _____
_____
_____

Is the millstone still attached to the parent bedrock or boulder? _____
If yes, describe: _____
_____

Appendix A

Can you tell why the millstone was rejected (material flaws, broken while trimming edges, undercut while shaping sides, broken while leveling top, broken while cutting the eye, broken while separating from the parent rock, etc.): _____
_____
_____

Does the central hole (eye) go completely through the millstone? ___ If not, how deep is the hole: ___ cm (___ inches).

Are there drill holes or tool marks on the millstone? If yes, describe: _____
_____
_____
_____

How far apart are the drill holes? 1 & 2: ___ cm, 2 & 3: ___ cm, 3 & 4: ___ cm, 4 & 5: ___ cm, 5 & 6: ___ cm, 6 & 7: ___ cm, 7 & 8: ___ cm, 8 & 9: ___ cm, 9 & 10: ___ cm, 10 & 11: ___ cm, 11 & 12: ___ cm, 12 & 13: ___ cm, 13 & 14: ___ cm, 14 & 15: ___ cm, 15 & 16: ___ cm,_____

What are the diameters of the drill holes? (key to a sketch map): 1) ___ cm, 2) ___ cm, 3) ___ cm, 4) ___ cm, 5) ___ cm, 6) ___ cm, 7) ___ cm, 8) ___ cm, 9) ___ cm, 10) ___ cm, 11) ___ cm, 12) ___ cm, 13) ___ cm, 14) ___ cm, 15) ___ cm, 16) ___ cm, 17) ___ cm, 18) ___ cm.

How deep are the drill holes? 1) ___ cm, 2) ___ cm, 3) ___ cm, 4) ___ cm, 5) ___ cm, 6) ___ cm, 7) ___ cm, 8) ___ cm, 9) ___ cm, 10) ___ cm, 11) ___ cm, 12) ___ cm, 13) ___ cm, 14) ___ cm, 15) ___ cm, 16) ___ cm, 17) ___ cm, 18) ___ cm.

Photographs

Black and white film:_____ (Roll) _____, Negative Numbers _____-_____
Color slide film: _____ (Roll) _____, Negative Numbers _____-_____
Location of photographs_____

Recorder and Date

Recorded by: _____

Address: _____
_____

Date recorded: _____

Other Comments

_____
_____
_____

Draw a detailed sketch of the millstone on the back of the form.

B. Form for Documenting Boulders and Drill Holes

MILLSTONE QUARRY DRILL HOLE RECORDING FORM

Location and General Data

Field Specimen Number: _____ Current Location: _____
_____

County: _____ State: _____

Setting of Specimen: ____ streambed, ____ quarry, ____ slope of hill, ____ top of hill, ___ base of hill, ___ other (_____).

Original location (if moved after original use): _____
_____

Owner: _____ Owner's Address: _____
_____

Archaeological Site Number: _____

U.S.G.S. 7.5 Minute Topographic Quadrangle: _____

Description

Raw Material: _____ conglomerate, _____ sandstone, _____ chert, _____ limestone, _____ other (_____)

Description of raw material including texture, inclusions, color, size and color of pebbles, etc.
_____
_____
_____

Describe the rock: ___ boulder, ___ quarried slab, ____ bedrock, ____ other (_____).

How large is this stone? Length: ___ cm (___ inches), Width: ___ cm (___ inches), and Thickness: ____ cm (____ inches).

How many drill holes are present? _____
The holes are: ____ cross-sectioned, ____ complete holes, or ____ base of complete hole. If more than one type of occurs, write the number of each hole type in the blanks above.

What type of fracture was produced when the stone was split? ___ straight, ___ concave, ___ convex, ____ irregular, ____ other (_____).

If more than one face of the stone has drill holes, please specify the type of break produced on each face: _____

Appendix B

Hole No.	Diameter		Length		Orientation	
	cm	inches	cm	inches	vertical	horizontal
1.	___	___	___	___	___	___
2.	___	___	___	___	___	___
3.	___	___	___	___	___	___
4.	___	___	___	___	___	___
5.	___	___	___	___	___	___
6.	___	___	___	___	___	___
7.	___	___	___	___	___	___
8.	___	___	___	___	___	___
9.	___	___	___	___	___	___
10.	___	___	___	___	___	___
11.	___	___	___	___	___	___
12.	___	___	___	___	___	___
13.	___	___	___	___	___	___
14.	___	___	___	___	___	___
15.	___	___	___	___	___	___
16.	___	___	___	___	___	___
17.	___	___	___	___	___	___
18.	___	___	___	___	___	___
19.	___	___	___	___	___	___
20.	___	___	___	___	___	___

How far apart are the drill holes? 1 & 2: ___ cm, 2 & 3: ___ cm, 3 & 4: ___ cm, 4 & 5: ___ cm, 5 & 6: ___ cm, 6 & 7: ___ cm, 7 & 8: ___ cm, 8 & 9: ___ cm, 9 & 10: ___ cm, 10 & 11: ___ cm, 11 & 12: ___ cm, 12 & 13: ___ cm, 13 & 14: ___ cm, 14 & 15: ___ cm, 15 & 16: ___ cm, 16 & 17: ___ cm, 17 & 18: ___ cm, 18 & 19: ___ cm, 19 & 20: ___ cm, _____

_____

Are there drill holes or tool marks on the stone? If yes, describe (length, width, depth, alignment, etc.): _____
_____
_____

Photographs

Black and white film: _____ (Roll) ____, Negative Numbers ____-____
Color slide film: _____ (Roll) ____, Negative Numbers ____-____
Location of photographs _____

Recorder and Date

Recorded by: _____

Address: _____
_____

Date recorded: _____

Other Comments

_____
_____
_____
_____
_____
_____
_____
_____

Draw a detailed sketch of the stone on the back of the form.

C. 1804 Lawsuit in Fayette County, Higbee v. Hanks

John Higbee versus Absolom Hanks concerning a pair of Red River millstones in 1804 filed in the Fayette County, Kentucky, Circuit Court. Source: Fayette County Circuit Court Records, original documents on file at Kentucky Department for Libraries and Archives, Frankfort, Kentucky.

Lexington District Court

    Fayette County Sct

    John Higbee by his attorney complains of Absolom Hanks of a plea of covenant broken; for that whereas the said Deft by his certain writing obligatory sealed with his seal and to the Court now here shewn dated of 10th Day of Dec. 1799 at the parish of Kentucky and County aforesaid did covenant and agree to and with the said Plff that he the said Deft would deliver at the dwelling house of the said Plff or at his Mill in the said County one pair of good Mill stones of the following dimensions to wit the running stone to be of the size of four feet diameter and twenty inches deep to be completely done of good and sufficient stone fitted for grinding corn in a compleat manner; the bed stone to be of the same diameter round and twelve inches thick to be of the same quality above mentioned and to be cut out of the stone quarry on red river in Montgomery County and to be delivered as above on or before the first day of January 1800 at the said Plffs dwelling house or Mill. And the said Plff agreed to pay the Deft for the said Mill stones when delivered as aforesaid fifteen pounds current money and one hundred and four gallons of merchantable whiskey in a good cask for the purpose — And the said Plff saith that altho he hath well and truly performed and kept all the covenants on his part to be performed and kept; yet the said Deft did not deliver on or before the first day of January 1800 or at any time since either at the said Plff dwelling house or Mill, the said Pair of Mill stones of the quality and description aforesaid; but he altogether neglected and refused to deliver the said Mill stones of the quality and description aforesaid altho the Plff was ready and willing to have received them — and so the Plff saith that the said Deft although often requested his covenant aforesaid hath not kept but hath broken to the Plff. Damage £100 and therefore he sues Brower for Plff.

    Know all men by these present that Absalom Hanks & a Radford McCorga — of Clark County & state of Kentucky are held & firmly bound unto Dillard Collins sheriff of the said County in the sum of two hundred pounds current money of Kentucky which payment well & truely to be made unto the said Dillard C. Allins his Heirs assigns &c we bind our selves our Heirs &c jointly severally & formerly by these presents sealed with our seals and dated this 19th day of Sept. 1800. [The next paragraph indicated that the Sheriff of Clark County arrested Absolom Hanks and he had to post bail before appearing in Fayette County Circuit Court.]

John Higbee v. Absolom Hanks

Case filed in 1804

Articles of agreement entered into and made this 10th day of Dec. 1799 Between John Higbee of the County of Fayette and state of Kentucky of the one part & Ab. Hanks of the County of Clark & state aforesaid of the other part Witnesseth that the said Ab. Hanks on his part doth Covenant & agree to and with the said John Higbee to Deliver at his Dwelling House, or at his mill in the sd county of Fayette a pair of Good Mill Stones of the dimensions to wit that is to say the said runing Stone to be of the size of four feet Diameter & twenty Inches Deep which is to be completely done of Good & sufficient stone fitted for Grinding Corn in a Compleat manner. The Bed Stone to be of the same Diameter round & twelve Inches thick to be of the same Quality as above mentioned & of the same Quality to be Cut out of the stone Quarry on Red River in this County or in the County of Montgomery & the above said stone be don & Delivered as above mentioned on or before the first day of Jany next ensuing at the said Higbee House or Mill and the said John Higbee on his part doth agree to and with the said Hanks to pay him the said Hanks and his Heirs Excrs admer assign the same of fifteen pounds Cur Muny & one Hundred & four Gallons of Good Merchantable Whiskey in a good Cask for the Purpose at the time of Delivery of the said Mill Stones. Wherefore we the said parties to this Article do agree to fix our hands & seals the day & date above written.

The document is signed by John Higbee and by A. Hanks. Witnessed by John Ireland. The court document is a true copy given the articles left in my hands by the Parties.

O. Young

test John Holloway

Robt. Nicholson

[Note: The Articles of Agreement are very faded and difficult to read.]

D. 1810 Lawsuit in Clark County, Wilkerson v. Adams

Joseph Wilkerson versus Spencer Adams concerning a pair of Red River millstones in 1810 filed in the Clark County, Kentucky, Circuit Court. Source: Clark County Circuit Court Records, original documents on file at Kentucky Department for Libraries and Archives, Frankfort, Kentucky.

Clarke Circuit Sct.

Joseph Wilkerson (assignee of William Palmer, who is assignee of Owen Dolly, who is assignee of Joel Tanner, who is assignee of Moses Treadway, who is assignee of James French) complains of Spencer Adams in a plea of Covenant broken, for that whereas the said defendant on the twenty sixth day of July in the year 1807 at the Circuit aforesaid, by his certain deed of covenant, sealed with the seal of the said defendant and to the Court now here shewn, the date whereof is the same day and year aforesaid, for value received bound himself, his heirs &c. to deliver to the aforesaid James French, his heirs or assigns one pair of millstones at the Red River Quarry in the County of Clarke one month after the same should be demanded the said mill stones to be of the best quality and of any size not exceeding three feet in diameter, which said French directed otherwise, and the Runner or upper millstone to be seventeen inches through the eye, if not directed otherwise, and the Bed stone of a corresponding thickness, and the said Stones to be cut in a neat and workmanlike manner: and whereas afterwards, to wit, on the fifth day of November 1807 at the Circuit aforesaid the said James French by an endorsement on said deed subscribed with the proper name of the said James, did assign the said deed to the aforesaid Moses Treadway (the said deed being then and there unperformed and in full force) of which said assignment the said defendant, then and there had due notice, by virtue whereof, pursuant to the Act of Assembly in such cases made an action accrued to the said Moses to demand and have of the defendant the aforesaid Millstones according to the form and effect of said deed. And whereas afterwards to wit on the seventh day of January 1808, at the Circuit aforesaid the said deed then and there being still unperformed and full force) the aforesaid Moses Treadway, by an assignment endorsed on said deed, subscribed with the proper name of the said Moses, transferred the said deed to Joel Tanner, of which the said defendant then and there had due notice whereby, pursuant to the Act of Assembly in such case provided an action accrued to the said Joel to demand and have of the said defendant the mill-stones aforesaid agreeably to the effect of said deed. And whereas afterwards, to wit on the eleventh day of January 1808 at the Circuit aforesaid (the said deed being then and there still unperformed and in full force) the said Joel Tanner, by an assignment endorsed on said deed, subscribed with the proper name of the said Joel, transferred the said deed to the aforesaid Owen Dolly for value received, of which the said defendant then and there had notice by reason whereof, pursuant to the Act of Assembly in such case made, an action accrued to the said Owen to demand and have of the said defendant the aforesaid mill-stones according to the effect of said deed: And whereas afterwards to wit, on the twenty ninth day of May 1808 at the Circuit aforesaid the said deed, then and there, being still unperformed and in full force) the said Owen Dolly, by an endorsement on said deed subscribed with the proper name of the said Owen, assigned said deed to the aforesaid William Palmer for value received of which the said defendant, then and there had due notice, by virtue whereof agreeably to the Act of Assembly in such case provided, an action accrued to the said William to demand and have of the said defendant the mill-stones aforesaid; And whereas afterwards, to wit on the eleventh day of November 1808 at the Circuit aforesaid, (the said deed being then and there still unperformed and in full force,) the said William Palmer, by an assignment endorsed on said deed transferred the same to the said Plff., which assignment was subscribed with the proper name of the said William, of which the said defendant, then and there had due notice, by virtue of which assignment, an action accrued to the said Plff. to demand and have of the said defendant the aforesaid mill-stones agreeably to the effect of said deed. And the said Plff. avers that he the said Plff., did on the day of in the year 1810 at the Circuit aforesaid, demand the aforesaid mill-stones of the said defendant; Nevertheless the said defendant, did not deliver to the said Plff. in one month after the date of the demand aforesaid, or at any time thereafter, one pair of mill-stones at the Red River Quarry in Clarke County agreeably to his deed of covenant aforesaid, of the size dimensions & quality described in said deed of covenant, (altho so to do the said defendant hath been often requested) and the said Plff. says that the said defendant hath not kept and performed his covenants aforesaid, but hath broken the same to the damage of the Plff. $200—wherefore he brings suit &c.—

[Among the other documents attached to the lawsuit was an agreement between James French and Spencer Adams:]

For the value received, I bind and oblige myself, my Heirs &c. to deliver to James French, his Heirs, or assigns one pair of mill stones at the Red River Quarry, in the County of Clarke, one month after the same shall be demanded, the mill stones to be of the best quality, and of any size not exceeding three feet in diameter, which said French directed otherwise, and the Runner or upper millstone to be seventeen inches through the eye, if not directed otherwise, and the Bed stone of a corresponding thickness, and the stones to be cut in a neat and workmanlike manner

As witness my hand and seal the 26th day of July 1807.

Spencer Adams Seal

Test—
Wm Calk

Hanson for Plff.

Wilkerson Assignee vs. Adams Decl. & Deed 1810 Sept. Cont defts costs & Dismd. Agreed.

E. 1823 Lawsuit in Clark County, Johnson v. Adams

Martin Johnson (Johnston) Administration versus Spencer Adams regarding Martin Johnson's work at the Red River millstone quarry during 1819 filed in the Clark County, Kentucky, Circuit Court in 1823. Source: Clark County Circuit Court Records, original documents on file at Kentucky Department for Libraries and Archives, Frankfort, Kentucky.

MARTIN JOHNSON

Account against Spencer Adams for work done in 1819 at the stone quarry in cutting of Millstones for which there is two hundred and thirty five dollars yet a balance due me.

CLARKE COUNTY & CIRCUIT SCT

William Johnson, administrator of all and singular the goods, chattels and credits which were of Martin Johnson deceased, at the time of his death, who died intestate, complains of Spencer Adams on a plea of Trespass on the case. For that whereas the said Spencer Adams on the __ day of __ in the year ____ during the life time of the said Martin Johnson at said County & Circ. & being indebted to him the said Martin Johnson in the sum of $235 for work and labour done and performed before the time by the said Martin Johnson, for him the said Spencer Adams at his the defendants special instance and request, in consideration thereof then and there promised the said Martin Johnson to pay him the said sum of $235 whenever required.

And whereas also afterwards to wit, on the same day and year aforesaid, at said County and Circuit during the life time of the said Martin Johnson, in consideration that the said Martin Johnson, at the special request of the said defendant, before that time had done and performed certain other work and labour, according to the account annexed for the benefit of the defendant, he the defendant, promised the said Martin Johnson to pay him therefore as much money as he reasonably deserved to have on demand. And the plaintiff avers that the said Martin Johnson reasonably deserved to have therefore the sum of $__ of which the defendant then and there had due notice.

Yet the said defendant not regarding his said promises, but contriving to defraud the said Martin Johnson in his life time, and the said William Johnson as administrator as aforesaid, after the death of the said Martin Johnson (to which said William Johnson after the death of the said Martin Johnson, to wit on the __ day of __ at the County & Circuit aforesaid administration of all and singular the goods chattels and credits, which were of the said Martin Johnson decd. at the time of his death who decd. intestate, in due form of law was granted, and the plaintiff brings here into Court a certificate of such grant of administration, whereby it will fully appear to the Court here that the plaintiff is administrator of the estate of said Martin Johnson, deceased) in this behalf. Hath not as yet paid that said sum of money or any part thereof to the said Martin Johnson in his lifetime or to the plaintiff administrator as aforesaid, since the death of the said Martin Johnson, although often requested so to do, but he so to do, hath hitherto wholly refused, and still refuses to pay the same or any part thereof to, to the said William Johnson administrator as aforesaid, to the damage of the plaintiff as administrator as aforesaid $300 therefore he sues &c.

Allan & Simpson Pf.

Filed 19th July 1821

[Note: The blanks in this copy were not filled in on the original handwritten version.]
JOHNSON ADMN. VS. ADAMS: INTERROGATION

Be it remembered that on the trial of this suit the plaintiff proved by a witness, Jacob Moore, that in the year 18__ , the defendant was the proprietor & occupant of a Mill Stone Quarry; & the witness, the plaintiffs intestate & a number of other labors at said quarry during said year. Witness commenced about 1st of April & worked constantly & regularly at said quarry until the last of November or 1st of December with the exception of the loss of a few weeks in Oct. & Nov. Witness found his own diet the balance of the hands were boarded by the defendant. He did not make himself any special contract with defendant nor does not know that there was any between said defendant & the other laborers, except that it seemed to be a general understanding amongst them that they were to have a proportionable part of the profits, to be estimated according to the length of time they worked. The plff's intestate commenced about the last of May or 1st of

June & continues in all about 3 months; after he had been there some time he heard deft. say he intended giving him (said intestate) a share according to the time he worked. Does not know how much was made by said deft., that or any other season at the quarry. Some time after the season was over deft. informed him that he could not get the hands together to make a settlement & he allowed the witness $100 for his services with permission to cut a pair of millstones & hand to assist him. Plaintiff's intestate was an old man but called a good hand at the business having had as he understood considerable experience tho' he was not superior to Witness in his judgement. Plaintiff's intestate had also a set of tools necessary to use in a quarry, which he worked with [**note**: this sentence had a horizontal line marked through each word]. Witness did not know what length of time plaintiff's intestate worked at said quarry, but supposed that it did not exceed three months in all; neither did he know exactly how long he had worked in said quarry himself during the aforesaid time he was working there occasionally. He did not know what was the contract between plaintiff's intestate and the defendant neither did he know whether there was any contract or not or upon what terms he worked for defendant.

DEPOSITION OF CORNELIUS SPRY

The deposition of Cornelius Spry taken at the Office of Allan & Simpson in the Town of Winchester on Saturday the 19th day of October 1822, in pursuance of an order of the Clarke Circuit Court to be read as evidence de bene esse in a suit at Common Law pending in the Clarke Circuit Court in which William Johnson as the administrator of his father Martin Johnston deceased is plaintiff and Spencer Adams is defendant.

Who being first sworn deposeth and sayeth: That he was acquainted with Martin Johnston deceased, that the year before said Johnston died, he commenced work with Spencer Adams about the first of the month of April in said year, and ended about the tenth day of November following. The work he was engaged at with said Adams was cutting millstones. He states that he was present at a conversation between said Martin Johnston and Spencer Adams after Johnston had quit work, which was the usual season of the year to leave off working in the stone quarry, an in said conservation, this deponent understood said Johnston and Adams to state that Johnston had worked with Adams, and was to get by the contract between them one fourth of what was made in that season out of the quarry after the deduction of expenses. And they also stated after having made some calculations that there had been made that season out of said quarry in all about fifteen hundred dollars, and about twelve hundred clear after the payment of all expenses. Adams was to collect all the money arising from the working of the quarry was to superintend the working of the same, had made all sales and was to pay over to said Johnston one-fourth part of the clear profits which I understood them in said conversation was as I have before stated the sum of twelve hundred dollars.

He states that he lived at that time with said Martin Johnston and that afterwards while said Johnston was confined to his bed by sickness and a short time before he died he sent this deponent to see said Spencer Adams and tell him he wanted to see him, but not to tell him what he wanted to see him for or he Adams might not come as he wanted to come to a settlement with him, and have his business arranged. Adams went to Johnstons house and Johnston said something about a settlement with him about the money that was coming to him from Adams, when he Adams replied that he had not his books along with him and therefore could not tell exactly how much it would be. Johnston replied that he could tell within ten dollars of it and observed that there was then coming to him from said Adams about the sum of three hundred dollars to which Adams replied that he was not far wrong.

This deponent further states that just as said Adams was leaving the house after said conversation had passed Johnston observed to him that he wanted him to come back shortly and bring his books with him in order that they might have a final settlement, as he was then sick and did not know whether or not he would recover which Adams consented to do and promised to return within a week. But did not come within the promised time and the old man Johnston then sent for his own son William Johnston and got him to draw off his account for him.

Being crossexamined by Defendant he says: He heard the conversation between Johnston & Adams at Johnstons House at the time Adams promised to return in about a week just as Spencer Adams was leaving there to go home. He was not present when the contract was made between Adams & Johnston. Johnston also stated that some person had got for him from Adams about $100 of what was coming from Adams about said work, but he does not recollect what the name of that person was. He does not remember of any person being present at these conversations. He does not know how long it takes to cut a pair of millstones or what a pair of Stones is worth. He does not know that all the work that was cut at said quarry during the time said Johnston worked there was sold. He does not know except from hearsay, that any of said work was ever sold. He does not recollect who stated the contract between said Adams & Johnston, more than he has already stated.

By Defendant: Did you understand from the conversation between Adams & Johnston — that they were equal partners in the stone cutting business, and that Johnston was to have one equal fourth part of all the profits, after a deduction of the expenses?

Answer: I understood that Johnston was to have one equal fourth part of all the profits after the expenses were deducted.

And further this deponent saith not.

Cornelius (X) Spry

(his mark)

Commonwealth of Kentucky Clarke County Sct. The foregoing deposition of Cornelius Spry was sworn to before me the subscriber a Justice of the peace for the county aforesaid at the time and place stated in the preamble thereto given under my hand this 19th day of October 1822.
Thos. G. Janies?

JOHNSTON ADMINISTRATION VS. ADAMS Bill of Exceptions

Filed 9 April 1823.

This being the whole of the evidence in this cause, & the Jury having found a verdict for the Plf f. for $50 in damages the deft. moves the Court to set the same aside and award a new trial on the grounds filed to wit: (Here we set them) which motion the Court overruled, to which opinion of the Court the defendant excepts and prays that his Bill of Exceptions may be signed, sealed & enrolled which is done accordingly.

Jas. Clark E.

F. 1826 Lawsuit in Clark County, Summers v. Adams

Cornelius Summers versus Spencer Adams Concerning a Pair of Red River Millstones in 1826 filed in the Clark County, Kentucky, Circuit Court. Source: Clark County Circuit Court Records, original documents on file at Kentucky Department for Libraries and Archives, Frankfort, Kentucky.

Clark County & Circuit

Cornelius Summers complains of Spencer Adams in custody & of plea of Covenant broken. For that whereas the said Deft. on the 15th day of June in the year 1825, at said Covenant, by his certain writing obligatory subscribed with his proper hand and sealed with his seal, which is to the Court now here produced, the date whereof, is the same day & aforesaid covenanted & promised to pay to the plaintiff eight months after the date of said writing one pair of four feet mill stones to be of the best quality & nicely cut & of usual thickness, & if they should not prove good, the Deft further covenanted to cut & deliver others in their place, till they should be good and the delivered at the Red River Stone Quarry for value need? and the plff avers, that on the 13th day of February in the year 1826 ... at said Red River Stone Quarry, & was there & ready & willing to receive from the Deft the said one pair of four feet mill stones of the best quality & nicely cut & of usual thickness- But that the Deft on that day failed to attend at said Quarry ... were send stones offered to him by any persons but at an Circuit? and the plff often saith, that the Deft his Covenant affords & to keep & performs, to ... the Deff, according to the true intent & weaning thereof, although often requested to do so & particularly as he said 15th day of February 1826 at said Stone Quarry as offered, hath not, but hath broken the same in this Covenant — that said Deft hath not paid to the plff, eight months after the date of said Covenant, at said Stone Quarry or any where else one pair of four feet mill stones of the best quality, & nicely cut & of usual thickness....

[Note: Lines drawn through some lines on this page. Also, appears that some text may be missing.]

...as by his said Covenant he ought to have ... but the said mill stones, to the plff to pay according to the form & effect of said covt. He (the Deft.) hath hitherto wholly failed & refused to do. & still doth refuse to the damages of the plff $200 & therefore he brings suit to

French & Farrow

[A second document in the lawsuit of Cornelius Summers against Spencer Adams is as follows:]

Clark County & Circuit

Cornelius Summers complains of Spencer Adams in custody & of plea of Covenant broken. For that whereas the said Deft. on the 15th day of June in the year 1825, at said Covenant, by his certain writing obligatory, subscribed with his proper hand and sealed with his seal, & which is to the Court now here produced, the date whereof, is the same day & year aforesaid, covenanted & promised to pay to the plaintiff on or before the first day of March next succeeding the date of said writing obligatory, eighty Seven dollars & fifty cents with Currency, for value vested? — and plff in fact saith, that the said Defendant by Covenant offers to keep & perform, to & with the plff, according to the true intent & weaning thereof, hath not, although often requested to do so but hath broken the same in this town, that the said Deft, the said eighty seven dollars fifty Cents in the Currency the said plff to pay, on or before the said first day of March next succeeding the date of said writing obligatory, hath not, or any part thereof, but hath the same as any part thereof, to the plff to pay in the Currency, or in any other thing, according to this form & effect of said Covenant, He (the Deft.) hath hitherto wholly failed & refused to do & still doth refuse, to the damages of the plff $100 & cheveton? He brings Suit to — French & Farrow

[The following agreement signed by Spencer Adams was included with the lawsuit:]

June the 15th of 1825
Eight months after date I promise to pay Cornelius Summers one pair of four feet mill stones to be of the best quality & nicely cut & of usual thickness & if they don't prove good then S Adams binds him self to cut & deliver others in the place till they are good to be delivered at the Red River Stone quarry writing my hand & seal this day & ... mark in

Spencer Adams

[One small piece of paper contained the verdict of the trial. It stated, "We of the jury find ... for the plantiff 150 dollars in damages. Isaac Conkwright."]

G. Millstones Potentially Associated with the Powell County Millstone Quarries

Several millstones were examined that appear to be associated with the Powell County millstones quarries. The author documented one of these specimens during a power line survey in October of 1991 (Hockensmith 2000). Also, four millstones in the Red River Historical Society and Museum's collection were found in the general vicinity of the quarries. These millstones were documented on March 6, 1988. Finally, several millstones were examined in a private collection in Bourbon County that may be from Pilot Knob. Oral history suggests that the original collector of these millstones removed some specimens from the Pilot Knob Quarry in the 1930s. Three of these millstones documented by the author on August 22, 1988, are included in this appendix.

Isolated Millstone

An isolated millstone was located on the forested north slope (960 feet above mean sea level) of an upland ridge system that connects Kit Point to the southwest and Rotten Point to the northeast

(Hockensmith 2000). The millstone was lying in the dry bed of a seasonal stream (ca. 1 meter wide) that feeds into a nearby intermittent stream. An examination of the millstone reveals that it is complete or nearly complete. It is made from a Pennsylvanian age conglomerate that has a light tan to light gray sandstone matrix containing rounded quartz pebbles (mostly white with a few yellowish brown). A large sandstone inclusion is present in the exposed face. This specimen is 92 cm in diameter and varies in thickness from 13 to 19 cm. The round eye (central hole), 18 to 19 cm in diameter, was completely cut through the stone. The millstone was rejected for obvious flaws. One edge has a break about 28 cm long that extends into the stone a maximum of 10 cm. A couple of smaller areas of edge damage are also present. The stone also has some undercutting on the side that makes the thickness vary. Leif Meadows and Rhondle Lee searched the area surrounding this millstone but found no other evidence of quarrying. About 150 m further west we observed a boulder in a streambed that had a possible shallow drill hole scar. It was immediately south of the Baker Millstone Quarry.

A couple of possible explanations can be offered for the presence of this isolated millstone. First, the millstone may have been manufactured from an isolated boulder which was a common manufacturing practice at the Powell County quarries (Hockensmith and Meadows 1996a, 1997). If this scenario is correct, the stone probably had some irregular breaks during the final shaping. Second, the stone may have gotten away from the millstone makers during transit. Sometimes millstones were rolled like a tire and got away from the workers (Hockensmith and Coy 1999:43). Such a stone could roll a considerable distance and be damaged. Edge damage would require cutting back the entire circumference to make a smaller stone. Since millstones were usually made to order, this was not an option in many cases. The presence of the millstone in this location is not surprising since it is roughly between the McGuire and Baker millstone quarries (Hockensmith and Meadows 1996a, 1997).

Red River Historical Society and Museums Collection

Four of the millstones documented in Red River Historical Society's collection during March of 1988 were originally located near the Powell County millstone quarries. These specimens were transported to the Red River Historical Society's museum for display to the public. Larry Meadows provided data on their original context. These specimens are described below.

Millstone # 1 was removed from road fill along Brush Creek which is located southwest of the Pilot Knob Quarry. This specimen was quarried from a light gray to light tan colored sandstone. It is 86 cm in diameter and is 45 cm thick. It has a round eye that is 17 cm in diameter. The millstone is roughly rounded but is extremely irregular in places. The exposed face is rough and uneven. It seems to have been abandoned due to shaping errors and a low area adjacent to the eye. Several faint horizontal tool marks in a radial pattern are visible on the upper face. One vertical drill hole 3 cm in diameter was observed as well. This specimen could either be a sandstone millstone or a very large grindstone.

Millstone # 3 was removed from Brush Creek about 300 yards below the forks of Brush Creek, southwest of the Pilot Knob Quarry. This specimen was made from a reddish brown conglomerate with very small quartz pebbles (most less than 5 mm in diameter). Only the Pilot Knob Quarry has pebbles as small as these. The millstone is 106 cm in diameter and is 30 cm thick. It has a round eye that is 18 cm in diameter cut through the stone. The millstone is rounded but not a perfect circle. The exposed face is also irregular. It seems to have been abandoned due to shaping errors that resulted in edge damage. An area of edge damage is 8 cm wide, 50 cm long, and 10 cm deep thus rendering the stone useless. Two parallel tool marks (9 cm apart) are visible on the upper face. One tool mark is 2.5 cm wide and 16 cm long while the other tool mark is 2 cm wide and 13 cm long.

Millstone # 4 was discovered southeast of the Lower McGuire Quarry, next to a little valley. This specimen was made from the typical conglomerate found in this area. It is slightly less than one-half of a complete millstone. The millstone is 43 cm in diameter and is 22 cm thick. It has a round eye that is 10 cm in diameter and 4 cm deep. Both the millstone and eye diameters would have been somewhat larger if they had been more complete. The millstone may have split apart when the eye was being cut. The exposed face is also irregular. It seems to have been abandoned due to shaping errors that resulted in edge damage.

Millstone # 8 was discovered during the excavations for a house on the lower west slope of Sage Point. This location is southwest of the Pilot Knob Quarry. This specimen was made from a reddish brown conglomerate with very small quartz pebbles (most less than 5 mm in diameter). Only the Pilot Knob Quarry has pebbles as small as these. The millstone is 141 cm in diameter and is 33 cm thick. It has a square eye that is 22 × 22 cm that extends through the stone. The millstone has been well rounded. The exposed face has an area of edge damage that is 6 cm wide, 48 cm long, and 5 cm deep. The face on the ground appears to be irregular. Thus, it appears to have been abandoned due to shaping errors. A horizontal drill hole scar is visible on one edge of the exposed face. This hole is 3 cm in diameter and 6 cm long. A vertical drill hole base is present on the exposed face 22 cm from the edge of the eye. This hole is 3 cm in diameter and 2 cm deep. There are also some possible tool marks on this millstone.

Private Millstone Collection

Oral history suggests that several millstones were removed from the Pilot Knob Quarry in the 1930s. Permission was obtained to document this Bourbon County millstone collection during August of 1988. The collection obviously contains specimens from several mills and some other contexts as well. While several millstones from this collection may have originated from the Pilot Knob Quarry, the three specimens described below are felt to be the most likely candidates since they are unfinished millstones.

Millstone # 18 is made from a light gray to light reddish brown conglomerate with very small quartz pebbles (most less than 5 mm in diameter). The millstone is 145 cm in diameter and is 30+ cm thick. It has a round eye that is 23 cm in diameter. The exposed portion of the millstone appears to be almost perfectly rounded with a flat upper surface. One edge of the upper surface has three parallel linear tree fossils and a rounded fossil elsewhere. A depression measuring 10 × 18 cm (5 cm deep) is present on one edge of the stone where an inclusion fell out. The fossils in this specimen would have made this a poor quality millstone.

Millstone # 19 is made from a light gray to a light reddish brown conglomerate with very small quartz pebbles (most less than 5 mm in diameter). The millstone is 132 cm in diameter and is 30+ cm thick. It has a round eye that is 22 cm in diameter. The millstone is well rounded but the edges are rough in a few places. The upper surface has been leveled but is not perfectly level. There is some minor edge damage as well.

Millstone # 21 is made from a light gray to a light reddish brown conglomerate with very small quartz pebbles (most less than 2 cm in diameter). The millstone is 135 cm in diameter and is 30+ cm thick. It has a roughed out rounded eye that is 26 cm in diameter. The millstone is roughly rounded and the upper surface is relatively level. There is one major area of edge damage that is 13 cm wide and 40 cm long and a smaller area of edge damage nearby. A groove 30 cm long extends from the major edge damage area to the edge of the eye. The groove is 4 cm wide and 2 cm deep.

Bibliography

Many of these sources have been briefly annotated, at least to the extent of identifying the subject or country being discussed. In some cases more detailed information has been provided, including translations of titles (placed in quotation marks).

Adams, R. Foster 1976. Some Clark County, Kentucky Adams Excerpts. In *Adams Addenda*, edited by Ruth Robinson and Dorothy Amburgey Griffith, pp. 25–26. St. Louis, Missouri.

Adams, William G. 1981. The Adams-Pigg Connection. Manuscript, 3 pp. Kentucky Heritage Council, Frankfort.

_____. 1989. Letter to Charles Hockensmith. On file at the Kentucky Heritage Council, Frankfort. [Discussing millstone maker Spencer Adams of Clark County, Kentucky]

Ambrose, Paul M. 1964. Abrasive Materials. In *Minerals Yearbook, 1963*, pp. 187–206. Department of the Interior, U.S. Bureau of Mines, Government Printing Office, Washington, D.C.

American Miller 1885. Hazards of Millstone Pickings. *American Miller* 13:25, January 1, 1885.

Anonymous n.d.a. Summers Family. Information in vertical files at Kentucky Historical Society Library, Frankfort.

Anonymous n.d.b. *Clark County Kentucky Marriages 1793–1850*. Researchers Publication.

Anonymous 1896. *Commemorative Biographical Encyclopedia of Dauphin County, Pennsylvania: Containing Sketches of Prominent and Representative Citizens, and Many of the Early Scotch-Irish and German Settlers*. J. M. Runk, Chambersburg, Pennsylvania.

Anonymous 1962. Stone-Age Skill Cut Millstones from Mountain: Boulders Scattered Over 50 Acres in Northeast Lancaster County Show Stonecutters' Work. *The Sunday News*, June 3, 1962, page 11. Newspaper clipping in vertical file at Lancaster County Historical Society, Lancaster, Pennsylvania.

Apperson, Richard, and James M. Bullock 1843. Report of the Commissioners of the Sinking Fund, in Relation to the Penitentiary. In *Reports Communicated to Both Branches of the Legislature of Kentucky, at the December Session, 1842*, pp. 385–399. A. G. Dodge, Frankfort.

The Argus of Western America 1812. Ad for Samuel Taylor's millstone quarry in Rockcastle County, Kentucky. *The Argus of Western America*, January 8, Frankfort, Kentucky.

_____ 1813. Ad for Charles Colyer's millstone quarry in Rockcastle County, Kentucky. *The Argus of Western America*, October 9, Frankfort, Kentucky.

_____ 1821a. Ad for Miller, Railsback & Miller's millstone quarry in Franklin County, Kentucky. *The Argus of Western America*, August 9, Frankfort, Kentucky.

_____ 1821b. Ad for Jeremiah Buckley's millstone quarry in Franklin County, Kentucky. *The Argus of Western America*, November 8, Frankfort, Kentucky.

Arndt, Karl J. R. 1975. *A Documentary History of the Indiana Decade of the Harmony Society 1813–1824*. Volume 1, 1814–1819. Indiana Historical Society, Indianapolis.

Arnow, Harriette S. 1983. *Seedtime on the Cumberland*. University Press of Kentucky, Lexington.

_____. 1984. *Flowering of the Cumberland*. University Press of Kentucky, Lexington.

Baber, Adin 1959. Nancy Hanks of "Undistinguished Families — Second Families." Privately printed, Bloomington, Indiana.

_____ 2004. *The Hanks Family of Virginia and Westward*. Revised edition compiled by Nancy Baber McNeill and Louis Franklin Hanks. Privately printed, Carpinteria, California.

Bailey, Harry H., and Joseph H. Winsor 1964. *Kentucky Soils*. Agricultural Experiment Station, University of Kentucky, Lexington.

Baker, Ira O. 1889. *A Treatise on Masonary Construction*. John Wiley, New York.

Ball, Donald B., and Charles D. Hockensmith 2005. Early Nineteenth Century Millstone Production in Tennessee. *Ohio Valley Historical Archaeology* 20:1–15.

_____, and _____ 2007a. Preliminary Directory of Millstone Makers in the Eastern United States. In *Millstone Studies: Papers on Their Manufacture, Evolution, and Maintenance* by Donald B. Ball and Charles D. Hockensmith, pp. 1–98. Special Publication No. 1. Jointly published by the Symposium on Ohio Valley Historic Archaeology, Murray, Kentucky, and the Society for the Preservation of Old Mills, East Meredith, New York.

_____, and _____ 2007b. Occupational Disease and Work Place Hazards Associated with Millstone Making. In *Millstone Studies: Papers on Their Manufacture, Evolution, and Maintenance*, by Donald B. Ball and Charles D. Hockensmith, pp. 99–105. Special Publication No. 1. Jointly published by the Symposium on Ohio Valley Historic Archaeology, Murray, Kentucky, and the Society for the Preservation of Old Mills, East Meredith, New York.

Baltimore American 1822. News item concerning millstones quarried in Montgomery County, Virginia. *Baltimore American*, July 9, 1822.

Barnes & Pearsol 1869. *Directory of Lancaster County: Embracing a Full List of all the Adult Males and Heads of Families, ... and a Classified Business Directory ... 1869–70*. Pearsol & Geist, Printers, Lancaster, Pennsylvania.

Beach, L. M., and A. T. Coons 1922. Abrasive Materials. In *Mineral Resources of the United States, 1919*, pp. 381–386. Part II—Nonmetals. Department of the Interior, U.S. Bureau of Mines, Government Printing Office, Washington, D.C.

_____, and _____ 1923. Abrasive Materials. In *Mineral Resources of the United States, 1920*, pp. 155–159. Part II—Nonmetals. Department of the Interior, U.S. Bureau of Mines, Government Printing Office, Washington, D.C.

_____, and _____ 1924. Abrasive Materials. In *Mineral Resources of the United States, 1921*, pp. 15–18. Part II—Nonmetals. Department of the Interior, U.S. Bureau of Mines, Government Printing Office, Washington, D.C.

_____, and _____ 1925. Abrasive Materials. In *Mineral Resources of the United States, 1922*, pp. 221–225. Part II—Nonmetals. Department of the Interior, U.S. Bureau of Mines, Government Printing Office, Washington, D.C.

Belmont, Alain 2006. *La Pierre à Pain: Les Carrières de Meules de Moulins en France du Moyen Âge à la Revolution Industrielle*. Presses Universitaries de Grenoble, France. [Two volume study on the millstone quarries of France dating from the Middle Ages to the Industrial Revolution]

_____, and Charles D. Hockensmith 2006. Millstones, Querns, and Millstone Quarry Studies in France: A Bibliography. *International Molinology* 72:2–11, Watford, England.

Berg, Thomas N. 1986. A Sesquicentennial Story: Early Millstone Quarry in Tioga County. *Pennsylvania Geology* 17 (1):3–6.

Bowles, Oliver 1930. Abrasive Materials. In *Mineral Resources of the United States, 1928*, pp. 237–252. Part II—Nonmetals. Department of the Interior, U.S. Bureau of Mines, Government Printing Office, Washington, D.C.

_____ 1932. Abrasive Materials. In *Mineral Resources of the United States, 1929*, pp. 65–81. Part II—Nonmetals. Department of the Interior, U.S. Bureau of Mines, Government Printing Office, Washington, D.C.

_____ 1939. *The Stone Industries*. McGraw-Hill, New York. Second edition.

_____, and A. E. Davis 1934. Abrasive Materials. In *Minerals Yearbook, 1934*, pp. 889–906. Department of the Interior, U.S. Bureau of Mines, Government Printing Office, Washington, D.C.

Bowman, James F. 1963. The John Treadway Family of Montgomery County, Kentucky. Manuscript, 28 pp. Kentucky Historical Society Library, Frankfort.

Boyd, Hazel Mason n.d. Cemeteries—Montgomery County, Kentucky. Manuscript. Kentucky Historical Society Library, Frankfort.

_____ 1961. *Some Marriages in Montgomery County, Kentucky before 1864*. Compiled and edited by Emma Jane Walker and Virginia Wilson. Kentucky Society, Daughters of the American Revolution.

Braun, E. Lucy 1950. *Deciduous Forest of Eastern North America*. Blakiston, Philadelphia.

Brown, Bruce B. 1990. *John Deeter: Allegheny Mountain Pioneer*. Closson Press, Apollo, Pennsylvania.

_____ [Greencastle, Pennsylvania] 1991. Letter to Charles D. Hockensmith, May 25. On file at the Kentucky Heritage Council, Frankfort.

Brundage, Larry 1990. Sawing Through the Grit, Part II. *The Chronicle of the Early American Industries Association* 43 (3): 65–67.

Butler, Brian M. 1971. *Hoover–Beeson Rockshelter, 40Cn4, Cannon County, Tennessee*. Miscellaneous Paper No. 9. Tennessee Archaeological Society, Knoxville.

Campbell, Marius R., et al. 1925. *The Valley Coal Fields of Virginia*. Virginia Geological Survey, Bulletin Number 25, Charlottesville.

Chandler, Henry P., and Annie L. Marks 1954. Abrasive Materials. In *Minerals Yearbook, 1951*, pp. 111–127. Department of the Interior, U.S. Bureau of Mines, Government Printing Office, Washington, D.C.

_____, and _____ 1955. Abrasive Materials. In *Minerals Yearbook, 1952*, pp. 99–114. Department of the Interior, U.S. Bureau of Mines, Government Printing Office, Washington, D.C.

_____, and _____ 1956. Abrasive Materials. In *Minerals Yearbook, 1953*, pp. 127–142. Department of the Interior, U.S. Bureau of Mines, Government Printing Office, Washington, D.C.

_____, and Gertrude E. Tucker 1953. Abrasive Materials. In *Minerals Yearbook, 1950*, pp. 90–107. Department of the Interior, U.S. Bureau of Mines, Government Printing Office, Washington, D.C.

_____, and _____ 1958a. Abrasive Materials. In *Minerals Yearbook, 1954*, pp. 117–132. Department of the Interior, U.S. Bureau of Mines, Government Printing Office, Washington, D.C.

_____, and _____ 1958b. Abrasive Materials. In *Minerals Yearbook, 1955*, Volume 1 (Metals and Minerals), pp. 121–140. Department of the Interior, U.S. Bureau of Mines, Government Printing Office, Washington, D.C.

_____, and _____ 1958c. Abrasive Materials. In *Minerals Yearbook, 1956*, pp. 139–157. Department of the Interior, U.S. Bureau of Mines, Government Printing Office, Washington, D.C.

_____, and _____ 1958d. Abrasive Materials. In *Minerals Yearbook, 1957*, pp. 145–164. Department of the Interior, U.S. Bureau of Mines, Government Printing Office, Washington, D.C.

_____, and _____ 1959. Abrasive Materials. In *Minerals Yearbook, 1958*, pp. 129–145. Department of the Interior, U.S. Bureau of Mines, Government Printing Office, Washington, D.C.

_____, and _____ 1960. Abrasive Materials. In *Minerals Yearbook, 1959*, pp. 137–153. Department of the Interior, U.S. Bureau of Mines, Government Printing Office, Washington, D.C.

_____, and _____ 1961. Abrasive Materials. In *Minerals Yearbook, 1960*, pp. 145–163. Department of the Interior, U.S. Bureau of Mines, Government Printing Office, Washington, D.C.

Child, Hamilton 1871. *Gazetteer and Business Directory of Ulster County, N.Y. for 1871–2*. H. Child, Syracuse, New York.

Cinadr, Thomas J., and David S. Brose 1978. *Archaeological Excavations in Caesar's Creek Lake, Ohio: Section II, The Carr Mill Race Site (33Wa75)*. Report submitted to the National Park Service, Atlanta, by the Cleveland Museum of Natural History, Cleveland.

Clark, Victor S. 1929. *History of Manufactures in the United States, Volume 1: 1607–1860*. Carnegie Institution of Washington. McGraw-Hill, New York.

Clark County, Kentucky 1797. Law Suit Filed in Clark County, Kentucky, Circuit Court. Valentine Huff versus Peter DeWitt Concerning a Plea of Debt. April 1. Kentucky State Archives, Frankfort. [Mentions that Peter DeWitt was a millstone cutter by trade]

_____ 1803. Law Suit Filed in Clark County, Kentucky, Circuit Court. James Daniel versus Martin DeWitt Concerning a Plea of Covenant Broken over 25 Grindstones on November 16, 1801. Kentucky State Archives, Frankfort.

_____ 1810. Law Suit Fled in Clark County, Kentucky, Circuit Court. Joseph Wilkerson versus Spencer Adams Concerning a Plea of Covenant Broken over a Pair of Millstones at the Red River Quarry in 1807. Kentucky State Archives, Frankfort.

_____ 1822. Law Suit Filed in Clark County, Kentucky, Circuit Court. Martin Johnson Administration versus Spencer Adams Regarding Work Done by Martin Johnson at the Millstone Quarry [Red River Quarry] in 1819. Kentucky State Archives, Frankfort.

_____ 1822. Law Suit Filed in Clark County, Kentucky, Circuit Court. Spencer Adams versus Peter Dewitt Concerning a Plea of Trespass for Attacking His Character and Good Name. Kentucky State Archives, Frankfort.

_____ 1822. Law Suit Filed in Clark County, Kentucky, Circuit Court. Spencer Adams versus Thomas Berry Concerning a Plea of Trespass for Attacking His Character and Good Name. Kentucky State Archives, Frankfort.

_____ 1826. Law Suit Filed in Clark County, Kentucky, Circuit Court. Cornelius Summers versus Spencer Adams Concerning a Plea of Covenant Broken over a Pair of Millstones at the Red River Quarry in 1825. Kentucky State Archives, Frankfort.

Clay, R. Berle 1994. Brutus' Industrial Complex, Mill and Saw Mill. Manuscript. Kentucky Heritage Council, Frankfort. [Contains dated journal entries and other information about grist and saw mills owned by Brutus Clay on his farm near Paris in Bourbon County, Kentucky]

Clifford, J.D. 1926. Geological and Mineralogical Observations. *Tennessee Historical Magazine* 9 (4):275–278.

Clift, G. Glenn 1966. *"Second Census" of Kentucky—1800*. Genealogical Publishing, Baltimore.

Clouse, Jerry A. [Pennsylvania Historical and Museum Commission, Harrisburg] 1991. Letter to Charles D. Hockensmith, February 2. On file at the Kentucky Heritage Council, Frankfort.

Collins, Gabriel 1848. *Gabriel Collins' Louisville and New Albany Directory and Annual Advertiser for 1848*. G. H. Monsarrat, Louisville, Kentucky.

Collins, Lewis 1847. *Historical Sketches of Kentucky*. Lewis Collins, Maysville, and J. A. & U. P. James, Cincinnati.

_____, and Richard H. Collins 1874. *History of Kentucky*. Collins, Covington, Kentucky.

Colyer, Charles 1824. Contract with Sidney Payne Clay, February 28, 1824. Sidney Payne Clay Papers (A\C621a, folder 5) Special Collections, Filson Historical Society, Louisville, Kentucky. [Handwritten contract between Colyer, of Rockcastle County, and Clay, of Bourbon County, for the delivery of two millstones to Clay]

The Commonwealth 1834. H. L. Soper's ad for French Burr Millstones. *The Commonwealth*, June 24, Frankfort, Kentucky.

Conley, James F. 1962. *Geology and Mineral Resources of Monroe County, North Carolina*. Bulletin 76, North Carolina Department of Conservation and Development, Division of Mineral Resources, Raleigh.

Coons, A. T., and B. H. Stoddard 1929. Abrasive Materials. In *Mineral Resources of the United States, 1926*, pp. 245–253. Part II—Nonmetals. Department of the Interior, U.S. Bureau of Mines Government Printing Office, Washington, D.C.

_____, and _____ 1930. Abrasive Materials. In *Mineral Resources of the United States, 1927*, pp. 91–98. Part II — Nonmetals. Department of the Interior, U.S. Bureau of Mines, Government Printing Office, Washington, D.C.

Cooper, James D., and Gertrude E. Tucker 1962. Abrasive Materials. In *Minerals Yearbook, 1961*, pp. 215–231. Department of the Interior, U.S. Bureau of Mines, Government Printing Office, Washington, D.C.

_____, and _____ 1963. Abrasive Materials. In *Minerals Yearbook, 1962*, pp. 197–212. Department of the Interior, U.S. Bureau of Mines, Government Printing Office, Washington, D.C.

Couey, Ann Poindexter 1975. *Clark County, Kentucky 1850 Census and Mortality Schedules 1852–1861*. Privately printed, Winchester, Kentucky.

Craik, David 1870. *The Practical American Millwright and Miller*. H. C. Baird, Philadelphia.

_____ 1882. *The Practical American Millwright and Miller*. Henry Carey Baird, Philadelphia.

Crawford, Byron 1999. Stones Offer a Look into the Past. *The Courier-Journal*, May 7, Louisville, Kentucky. [Discussing the Powell County, Kentucky millstone quarries]

Cross, Alice 1996. Meet You at the Station(s). *The Accordian* 10 (2):6–10, Rochester, New York.

Cuming, F. 1810. Sketches of a Tour to the Western Country Through the States of Ohio and Kentucky; a Voyage Down the Ohio and Mississippi Rivers, and a Trip Through the Mississippi Territory, and Part of West Florida Commenced at Philadelphia in the Winter of 1807, and Concluded in 1809. Reprinted in *Early Western Travels 1748–1846*, edited by Reuben Gold Thwaites (1904), Volume 4, pp. 17–377. Arthur H. Clark, Cleveland.

Darton, N. H. 1894. Shawangunk Mountain. *National Geographic Magazine*, March 17.

Davis, A. E. 1935. Abrasive Materials. In *Minerals Yearbook, 1935*, pp. 995–1010. Department of the Interior, U.S. Bureau of Mines, Government Printing Office, Washington, D.C.

Day, David T. 1886. Abrasive Materials. In *Mineral Resources of the United States, Calendar Year 1885*, pp. 428–430. Part II — Nonmetals. U.S. Bureau of Mines, U.S. Geological Survey, Government Printing Office, Washington, D.C.

_____. 1888. Abrasive Materials. In *Mineral Resources of the United States, Calendar Year 1887*, pp. 552–553. Part II — Nonmetals. U.S. Bureau of Mines, U.S. Geological Survey, Government Printing Office, Washington, D.C.

_____. 1890. Abrasive Materials. In *Mineral Resources of the United States, Calendar Year 1888*, pp. 576–577. Part II — Nonmetals. U.S. Bureau of Mines, U.S. Geological Survey, Government Printing Office, Washington, D.C.

_____. 1892. Abrasive Materials. In *Mineral Resources of the United States, Calendar Years 1889–1890*, pp. 556–557. Part II — Nonmetals. U.S. Bureau of Mines, U.S. Geological Survey, Government Printing Office, Washington, D.C.

Dean, Lewis S. [Geological Survey of Alabama, Tuscaloosa] 1996. Letter to Charles D. Hockensmith, November 21. [Discussing the millstone industry in Alabama]

_____ (editor) 1995. *Michael Tuomey's Reports and Letters on the Geology of Alabama, 1847–1856*. Information Series 77. Geological Survey of Alabama, Tuscaloosa.

Dedrick, B. W. 1924. *Practical Milling*. National Miller, Chicago. Reprinted in 1989 by the Society for the Preservation of Old Mills, Manchester, Tennessee.

Dunn, Thelma M. 1996a. *Montgomery County, Kentucky Tax Records 1797–1799–1800*. Privately printed, Atoka, Tennessee.

_____ 1996b. *Montgomery County, Kentucky Tax Records 1806–1807–1808–1809–1810 and the 1810 U S. Census Record*. Privately printed, Atoka, Tennessee.

Dyche, Russell 1941. Library of Mountain Millstones. Manuscript. Levi Jackson State Park, London, Kentucky.

_____ 1950. *Mountain Life Museum of Levi Jackson Wilderness Road State Park, London, Ky*. Laurel County, Kentucky Information Series, No. 12. The Sentinel-Echo, London, Kentucky.

The Eagle 1825a. Ad for Burr Millstones and carding machine by January and Sutherland of Maysville. *The Eagle*, February 9, Maysville, Kentucky.

_____ 1825b. Ad for Laurel-Hill Millstones from Pennsylvania. *The Eagle*, April 13, Maysville, Kentucky.

_____ 1825c. Ad for sale of water grist mill. *The Eagle*, September 14, Maysville, Kentucky.

_____ 1826. Ad for sale of steam grist mill in Maysville. *The Eagle*, March 26, Maysville, Kentucky.

Ebert, Milford 1993. On the Right Track with the Accord Train Station. *The Accordian* 7 (3):1–5, Rochester, New York.

Edwards, Richard 1855. *Statistical Gazetteer of the State of Virginia, Embracing Important Topographical and Historical Information from Recent and Original Sources, Together with the Results of the Last Census Population, in Most Cases, to 1854*. Richard Edwards, Richmond.

Emmons, Ebenezer 1852. *Report of Professor Emmons, on His Geological Survey of North Carolina*. S. Gales, Raleigh.

_____ 1856. *Geological Report of the Midland Counties of North Carolina*. Putnam, New York.

Enoch, Harry G., and Larry G. Meadows 2005. *Clark County Road Book: Index to Roads, Mills and Ferries in Clark County Order Books 1793–1876*. Red River Historical Society & Museum, Clay City, Kentucky.

Fayette County, Kentucky 1804. Law Suit Filed in

Fayette County, Kentucky, Circuit Court. John Higbee versus Absolom Hanks Concerning a Plea of Covenant Broken over a Pair of Millstones to Be Made at the Red River Quarry According to an Agreement on December 10, 1799. Kentucky State Archives, Frankfort.

Fedders, James M. 1983. *The Vegetation and Its Relationships with Selected Soil and Site Factors of the Spencer-Morton Preserve, Powell County, Kentucky.* Unpublished Master's thesis, Department of Biological Sciences, Eastern Kentucky University, Richmond.

Felldin, Jeanne Robey, and Gloria Kay Vandiver Inman 1981. *Index to the 1820 Census of Kentucky.* Genealogical Publishing, Baltimore.

Flick, Alexander C. 1927. *The Papers of Sir William Johnson.* Volume V. University of the State of New York, Albany.

_____ 1931. *The Papers of Sir William Johnson.* Volume VII. University of the State of New York, Albany.

_____ 1933. *The Papers of Sir William Johnson.* Volume VIII. University of the State of New York, Albany.

Flint, James 1822. Letters from America, Containing Observations on the Climate and Agriculture of the Western States, the Manners of the People, the Prospects of Emigrants, &c., &c. (1822) [For Years 1818–1820]. Reprinted in *Early Western Travels 1748–1846,* edited by Reuben Gold Thwaites, Volume 9, pp. 15–333. Arthur H. Clark, Cleveland.

Flory, Paul B. 1951a. Old Millstones. *Papers Read Before the Lancaster County Historical Society,* 55 (3):73–86, Lancaster, Pennsylvania.

_____ 1951b. Millstones and Their Varied Usage. *Papers Read Before the Lancaster County Historical Society,* 55 (5):125–136, Lancaster, Pennsylvania.

Fontaine, William M. 1869. The Building Stone and Slate of Virginia. *The Manufacturer and Builder,* Volume 1, Issue 2, pp. 46–47, Western, New York.

The Frankfort Commonwealth 1845. Ad for E. and S. Stedman's millstones in Franklin County. *The Frankfort Commonwealth,* March 11, Frankfort, Kentucky.

Fries, Robert 1995. Handmade, from the Last of the Stonecutters: Recollections of the Brush Mountain Quarries. *New River Current* 7 (300):16–18. *Roanoke Times & World-News,* February 5, Roanoke, Virginia. [Story about millstone makers Robert Houston Surface and W. C. Saville.]

_____ 1995. Millstone Memories: All that Remains of the Brush Mountain Quarries are the Recollections of Two Montgomery County Men, the Last of the Stonecutters Who Made Millstones by Hand. *Roanoke Times & World-News* (New River Valley Edition), February 5, 1995, page 16, Roanoke, Virginia. On NewsBank and ProQuest websites. [Story about millstone makers Robert Houston Surface and W. C. Saville]

Gannett, Henry 1883. Abrasive Materials. In *Mineral Resources of the United States, Calendar Year 1882,* pp. 476–481. Part II — Nonmetals. Department of the Interior, U.S. Bureau of Mines, Government Printing Office, Washington, D.C.

Garber, D. W. 1970. *Waterwheels and Millstones: A History of Ohio Gristmills and Milling.* Ohio Historical Society, Columbus.

Gary, Margaret, Robert McAlfee, Jr., and Carol L. Wolf (editors) 1974. *Glossary of Geology.* American Geological Institute, Washington, D.C.

Gilpin, Joshua 1926. Journal of a Tour from Philadelphia Thro' the Western Counties of Pennsylvania in the Months of September and October, 1809. *The Pennsylvania Magazine of History and Biography* (1926) 50:64–78, 163–178, 380–382; (1927) 51:172–190, 351–375; and (1928) 52:29–58.

Grassi, Robert 2004a. The Miller and Millstones — Part I. *Old Mill News* 32 (1):12–13.

_____ 2004b. The Miller and Millstones — Part II. *Old Mill News* 32 (2):16–17.

Grimshaw, Robert 1882. *The Miller, Millwright, and Millfurnisher: A Practical Treatise.* H. Lockwood, New York.

Halcomb, Clarence [Hamilton, Ohio] 1992. Letter to Charles D. Hockensmith, June 7. On file at the Kentucky Heritage Council, Frankfort.

Harms, Eduard, and Fritz Mangartz 2002. *Vom Magma zum Mühlstein: Eine Zeitreise durch die Lavaströme des Bellerberg-Vulkans.* Römisch-Germanisches Zentralmuseum, Vulkanpark-Forschungen, Band 5, Mainz, Germany. [Book that deals with the basalt millstone industry in the vicinity of Mayen in northwestern Germany]

Hart, Thomas, Jr. 1807. For Sale, A Merchant Mill, Saw Mill, Distillery, and Fifty Acres of Land. *Kentucky Gazette and General Advertiser* Vol. 20, No. 1138.

Hartnagel, C. A. 1927. *Mining and Quarry Industries of New York 1919–1924.* New York State Museum Bulletin Number 273, Albany.

Hartnagel, C. A., and John G. Broughton 1951. *Mining and Quarry Industries of New York 1937–1948.* New York State Museum Bulletin Number 343, Albany.

Harvey, H. H. 1884. *H. H. Harvey's Illustrated Catalogue and Price List for 1884 & 1885 of Stone Cutters', Quarrymen's, Miners', Railroad, Grist Mill, and Blacksmiths' Hammers, Sledges, and Tools, Sleds, &c., &c.* Kennebec Journal Printers, Augusta, Maine.

_____ 1886. *H. H. Harvey's Illustrated Catalogue and Price List for 1886 of Stone Cutters', Quarrymen's, Miners', Railroad, Grist Mill, and Blacksmiths' Hammers, Sledges, and Tools, Sleds, &c., &c.* Kennebec Journal Printers, Augusta, Maine.

_____ 1896. *H. H. Harvey's Illustrated Catalogue and Price List for 1896–1897 of Stone Cutters', Quarrymen's, Miners', Railroad, Grist Mill, Coopers', Blacksmiths', and Slaters' Hammers, Sledges, Tools, and Outfits; Also, Contractors' Supplies, Handles, Iron,*

Steel, etc., etc. H. H. Harvey, Augusta, Maine. Reprinted in 1973 by Early American Industries Association.

Hatmaker, Paul, and A. E. Davis 1932. Abrasive Materials. In *Mineral Resources of the United States, 1930*, pp. 151–169. Part II — Nonmetals. Department of the Interior, U.S. Bureau of Mines, Government Printing Office, Washington, D.C.

_____, and _____ 1933a. Abrasive Materials. In *Mineral Resources of the United States, 1931*, pp. 111–130. Part II — Nonmetals. Department of the Interior, U.S. Bureau of Mines, Government Printing Office, Washington, D.C.

_____, and _____ 1933b. Abrasive Materials. In *Minerals Yearbook, 1932–33*, pp. 647–667. Part II — Nonmetals. Department of the Interior, U.S. Bureau of Mines, Government Printing Office, Washington, D.C.

Hazen, Theodore R. 1996a. Millstone Dressing — Part 1. *Old Mill News* 25 (1):16–17.

_____ 1996b. Millstone Dressing — Part 2. *Old Mill News* 25 (3):8–9, 11.

Hehnly & Wike 1880. Broadside advertisement for Turkey Hill millstones. Hehnly & Wike, Durlach, Pennsylvania. Hagley Museum and Library, Wilmington, Delaware.

Heinemann, Charles B. 1934. *Daniel Families of the Southern States: A Compilation Covering One Hundred Thirteen Daniel Families.* Privately printed, Washington, D.C.

_____ 1976. *"First Census" of Kentucky 1790.* Genealogical Publishing Company, Baltimore.

_____, and Gaius M. Brumbaugh 1938. *First Census of Kentucky, 1790.* G. M. Brumbaugh, Washington, D.C.

Helton, Walter L. 1964. *Kentucky's Rocks and Minerals.* Series X, Special Publication 9. Kentucky Geological Survey, Lexington.

Hensley, Mrs. Chuckie Hall 1986. *Powell County, Kentucky 1910 Census.* McDowell Publications, Utica, Kentucky.

Heverly, Clement F. 1915. *Pioneer and Patriot Families of Bradford County, Pennsylvania.* Volume 2. Bradford Star Printer, Towanda, Pennsylvania.

Hirsch, Steve [Kingston, New York] 2005. E-mails to Charles D. Hockensmith, January 2 and 10, 2005. [Commenting on the Ulster County, New York, millstone quarries]

Hockensmith, Charles D. 1983. Letter to Fred E. Coy, Jr., September 19. On file at the Kentucky Heritage Council, Frankfort. [Discussing petroglyphs, millstone quarries, and saltpeter sites in Kentucky]

_____ 1988. The Powell County Millstone Quarries. *Kentucky Archaeology Newsletter* 6 (1):4.

_____ 1990a. The Ware Millstone Quarry. *Kentucky Archaeology Newsletter* 7 (1):5.

_____ 1990b. The Lower McGuire and Ewen Millstone Quarries. *Kentucky Archaeology Newsletter* 7 (2):5–6.

_____ 1990c. Interview with Virginia's Last Millstone Makers. *Kentucky Archaeology Newsletter* 7 (2):6.

_____ 1993a. *The Pilot Knob Millstone Quarry: A Self-Guided Trail.* Pilot Knob State Nature Preserve, Powell County, Kentucky. Booklet (limited edition) published by the Kentucky Nature Preserves Commission and the Kentucky Heritage Council, Frankfort.

_____ 1993b. Millstone Quarrying in the Eastern United States: A Preliminary Overview. *Ohio Valley Historical Archaeology* 7 & 8: 83–89.

_____ 1993c. Study of American Millstone Quarries. *Old Mill News* 21 (1):5–7; 21 (2):4–8.

_____ 1994a. *The Pilot Knob Millstone Quarry: A Self-Guided Trail.* Pilot Knob State Nature Preserve, Powell County, Kentucky. Booklet (second edition with run of 5,000 copies) published by the Kentucky Nature Preserves Commission and the Kentucky Heritage Council, Frankfort.

_____ 1994b. Millstone Quarrying and Pine Tar Manufacture: Two Historic Powell County, Kentucky Industries. Paper presented at the Red River Historical Society meeting at Natural Bridge State Park, Slade, Kentucky.

_____ 1999a. The Millstone Industry in Southwest Virginia. In *Millstone Manufacture in Virginia: Interviews with the Last Two Brush Mountain Millstone Makers,* edited by Charles D. Hockensmith, pp. 1–3. Society for the Preservation of Old Mills, Newton, North Carolina.

_____ 1999b. The Manufacture of Conglomerate Millstones at the Powell County, Kentucky Quarries. Paper presented at the Annual Meeting of the Society for the Preservation of Old Mills, Newton, North Carolina.

_____ 2000. *An Archaeological Reconnaissance of a 6.4 Mile Long Segment of the Proposed Smith to Stanton Power Line in Powell County, Kentucky.* Draft report for the Kentucky Heritage Council Occasional Reports in Archaeology, Number 3, Frankfort.

_____ 2002. The Conglomerate Millstone Industry in the Eastern United States. Paper presented at the Colloque International "Extraction, Façonnage, Commerce et Utilisation des Meules de Moulin — Une Industrie dans la Longue Durée," La Ferté-sous-Jouarre, France. May 16–19, 2002.

_____ 2003a. The Conglomerate Millstone Industry in the Eastern United States. In *Meules à Grains: Actes du Colloque International de La Ferté-sous-Jouarre 16–19 Mai 2002,* edited by Mouette Barboff, François Sigaut, Cozette Griffin-Kremer, and Robert Kremer, pp. 197–216. Éditions Ibis Press and Éditions de la Maison des Sciences de l'Homme, Paris, France.

_____ 2003b. The Millstone Industry in Kentucky: Brief Glimpses from Archival Sources. *The Millstone* 2 (1): 6–17. Kentucky Old Mill Association, Clay City, Kentucky.

_____ 2003c. The Ohio Buhr Millstones: The Flint

Ridge and Raccoon Creek Quarries. *Ohio Valley Historical Archaeology* 18:135–142.

_____ 2004a. *Early American Documents and References Pertaining to Millstones: 1628–1829*. Kentucky Old Mill Association, Clay City, Kentucky.

_____ 2005. The Preservation, Ownership, and Interpretation of American Millstone Quarries. Paper presented at the Colloque International "Les Meulières, Recherche. Protection et Valorisation d'un Patrimoine Industriel Européen (Antiquité-XXIe s.)," Grenoble, France, September 22–25, 2005.

_____ 2006. The Preservation, Ownership, and Interpretation of American Millstone Quarries. In *Mühlsteinbrüche. Erforschung, Schutz und Inwertsetzung eines Kulturerbes europäischer Industrie (Antike–21. Jahrhundert)*, edited by Alain Belmont and Fritz Mangartz, pp. 193–204. Actes du Colloque International de Grenoble, 22–25/9/2005, Römisch-Germanisches Zentralmuseum, Tagungen Band 2, Mainz, Germany.

_____ 2007. The Ohio Buhr Millstones: The Flint Ridge and Raccoon Creek Quarries. In *Millstone Studies: Papers on Their Manufacture, Evolution, and Maintenance*, edited by Donald B. Ball and Charles D. Hockensmith, pp. 134–143. Special Publication No. 1. Jointly published by the Symposium on Ohio Valley Historic Archaeology, Murray, Kentucky, and the Society for the Preservation of Old Mills, East Meredith, New York.

_____ 2008a. *The Millstone Industry: A Summary of Research on Quarries and Producers in the United States, Europe and Elsewhere*. McFarland, Jefferson, North Carolina. In press.

_____ 2008b. The Millstone Industry in New York. In *The Historic Millstone Industry in New York State with an Emphasis on Ulster County*, edited by Charles D. Hockensmith, Society for the Preservation of Old Mills. Draft manuscript.

_____ 2008c. The French Burr Millstone in Kentucky: Insights from Early Ads, 1792–1890. In *Foreign and Domestic Millstones Used in Kentucky: Papers Examining Archival Records*, compiled by Charles D. Hockensmith, pp. 5–38. Kentucky Old Mill Association, Clay City, Kentucky. In press.

_____ 2008d. Millstones from Ohio and Pennsylvania Imported into Kentucky: Raccoon Buhrs and Laurel Hill Stones. In *Foreign and Domestic Millstones Used in Kentucky: Papers Examining Archival Records*, compiled by Charles D. Hockensmith, pp. 39–54. Kentucky Old Mill Association, Clay City, Kentucky. In Press.

_____ 2008e. Early References to Red River Millstones in Kentucky: Newspaper Ads, 1803–1839. In *Foreign and Domestic Millstones Used in Kentucky: Papers Examining Archival Records*, compiled by Charles D. Hockensmith, pp. 55–62. Kentucky Old Mill Association, Clay City, Kentucky. In Press.

_____ (editor) 1999. *Millstone Manufacture in Virginia: Interviews with the Last Two Brush Mountain Millstone Makers*. Society for the Preservation of Old Mills. Marblehead Publishing, Raleigh, North Carolina.

_____ (editor) 2008a. *Foreign and Domestic Millstones Used in Kentucky: Papers Examining Archival Records*. Kentucky Old Mill Association, Clay City, Kentucky. Kentucky Old Mill Association, Clay City, Kentucky. In press.

_____ (editor) 2008b. *The Historic Millstone Industry in New York State with an Emphasis on Ulster County*. Society for the Preservation of Old Mills. Draft manuscript.

_____, and Fred E. Coy, Jr. 1999. Early Twentieth Century Millstone Manufacture in Southwest Virginia: An Interview with Millstone Makers Robert Huston Surface and W. C. Saville. In *Millstone Manufacture in Virginia: Interviews with the Last Two Brush Mountain Millstone Makers*, edited by Charles D. Hockensmith, pp. 5–62. Society for the Preservation of Old Mills. Marblehead Publishing, Raleigh, North Carolina.

_____, and _____ 2008a. Twentieth Century Conglomerate Millstone Manufacture Near Accord in Ulster County, New York: An Interview with Vincent and Wallace Lawrence. In *The Historic Millstone Industry in New York State with an Emphasis on Ulster County*, edited by Charles D. Hockensmith. Society for the Preservation of Old Mills. Draft manuscript.

_____, and _____ 2008b. The Esopus Millstone Industry at Accord, Ulster County, New York: An Interview with Lewis Waruch. In *The Historic Millstone Industry in New York State with an Emphasis on Ulster County*, edited by Charles D. Hockensmith. Society for the Preservation of Old Mills. Draft manuscript.

_____, and Larry G. Meadows 1996. Historic Millstone Quarrying in Powell County, Kentucky. *Ohio Valley Historical Archaeology* 11:95–104.

_____, and _____ 1997. Conglomerate Millstone Quarrying in the Knobs Region of Powell County, Kentucky. *Old Mill News* 25 (2):17–20; 25 (3):24–26.

_____, and _____ 2006. Red River Millstone Quarries in Lawsuits. Part I. *The Millstone* 5 (2):9–17. Kentucky Old Mill Association, Clay City, Kentucky.

_____, and _____ 2007. Red River Millstone Quarries in Lawsuits. Part 2. *The Millstone* 6 (1):31–38. Kentucky Old Mill Association, Clay City, Kentucky.

_____, and Jimmie L. Price 1999. Conglomerate Millstone Making in Southwestern Virginia: An Indepth Interview with Millstone Maker Robert Huston Surface. In *Millstone Manufacture in Virginia: Interviews with the Last Two Brush Mountain Millstone Makers*, edited by Charles D. Hockensmith, pp. 63–89. Society for the Preservation of Old Mills. Marblehead Publishing, Raleigh, North Carolina.

Hörter, Fridolin 1994. *Getreidereiben und Mühlsteine aus der Eifel* ["Querns and Millstones from the Eifel"]. Geschichte und Altertumsverein fur Mayen un Umgebung, Mayen. [Germany]

Howell, Charles 1985. Colonial Watermills in the Wooden Age. In *America's Wooden Age: Aspects of Its Early Technology*, edited by Brooke Hindle, pp. 120–159. Sleepy Hollow Press, Tarrytown, New York.

_____ 1993. International Mini-Mill Symposium. *Old Mill News* 21 (1):15–17.

_____ 1997. Millstones: An Introduction. *Old Mill News* 25 (4):18–22.

_____, and Allan Keller 1977. *The Mill at Philipsburg Manor and a Brief History of Milling*. Sleepy Hollow Restorations, Tarrytown, New York.

Hubble, Anna Joy Munday 1992. *Clark Co KY 1810*. Privately printed, Whitefish, Montana.

Hughes, William C. 1869. *The American Miller, and Millwright's Assistant*. Henry C. Baird, Philadelphia.

Irwin, John Rice [Museum of Appalachia, Norris, Tennessee] 1993. Personal communication to Charles D. Hockensmith. [Discussing a conglomerate millstone quarry in Union County, Tennessee]

Ison, Cecil R. 2005. Occupational Hazards of Milling. *The Millstone* 4 (1): 11–13. Kentucky Old Mill Association, Clay City, Kentucky.

Jackson, Ronald V. 1988a. *Kentucky 1860 East*. Accelerated Indexing International, Inc., North Salt Lake, Utah.

_____ 1988b. *Kentucky 1870 Federal Census Index—East*. Accelerated Indexing International, Inc., North Salt Lake, Utah.

_____, and David Schaefermeyer 1976. *Kentucky 1850 Census Index*. Accelerated Indexing Systems, Bountiful, Utah.

_____, and Gary R. Teeples 1976. *Kentucky 1820 Census Index*. Accelerated Indexing Systems, Bountiful, Utah.

_____, and _____ 1978a. *Kentucky 1810 Census Index*. Accelerated Indexing Systems, Bountiful, Utah.

_____, and _____ 1978b. *Kentucky 1830 Census Index*. Accelerated Indexing Systems, Salt Lake City, Utah.

_____, and _____ 1978c. *Kentucky 1840 Census Index*. Accelerated Indexing Systems, Bountiful, Utah.

_____, _____, and David Schaefermeyer 1976. *Kentucky 1830 Census Index*. Accelerated Indexing Systems, Bountiful, Utah.

Jagailloux, Serge 2002. État de Santé des Meuliers, Accidents du Travail, Affections Professionnelles. In *Les Meuliers. Meules et Pierres Meulières* by Agapain, pp. 151–164. Presses du Village, Étrépilly, France. [France]

Jegli, John B. 1850. *The Louisville Business Register for 1850*. Printed by Brennan & Smith, Louisville, Kentucky.

Jobey, George 1986. Millstones and Millstone Quarries in Northumberland. *Archaeologia Aeliana* 5th series, 14:49–80. [England]

Johnson, Bertrand L., and A. E. Davis 1936. Abrasive Materials. In *Minerals Yearbook, 1936*, pp. 877–894. Department of the Interior, U.S. Bureau of Mines, Government Printing Office, Washington, D.C.

_____, and _____ 1937. Abrasive Materials. In *Minerals Yearbook, 1937*, pp. 1283–1300. Department of the Interior, U.S. Bureau of Mines, Government Printing Office, Washington, D.C.

_____, and _____ 1938. Abrasive Materials. In *Minerals Yearbook, 1938*, pp. 1135–1150. Department of the Interior, U.S. Bureau of Mines, Government Printing Office, Washington, D.C.

_____, and M. Schauble 1939. Abrasive Materials. In *Minerals Yearbook, 1939*, pp. 1225–1240. Department of the Interior, U.S. Bureau of Mines, Government Printing Office, Washington, D.C.

Johnson, Joseph Risk 1965. Transcribed version of William Risk's interview with John Shane (John Shane, Draper MSS. 11CC 86). Copy on file at the Kentucky Historical Society, Frankfort.

Jones, Walter B. 1926. *Index to the Mineral Resources of Alabama*. Bulletin No. 28. Geological Survey of Alabama, University, Alabama.

Judd, William [John's Island, South Carolina] 1999. Letter to Charles D. Hockensmith, December 14. On file at the Kentucky Heritage Council, Frankfort. [Discussing the use of Peak Millstones in the Rice Industry in South Carolina]

_____ [John's Island, South Carolina] 2000. Letter to Charles D. Hockensmith, January 24. On file at the Kentucky Heritage Council, Frankfort. [Discussing the use of Peak Millstones in the Rice Industry in South Carolina]

Katz, Frank J. 1913. Abrasive Materials. In *Mineral Resources of the United States, Calendar Year 1912*, pp. 819–831. Part II—Nonmetals. Department of the Interior, U.S. Bureau of Mines, Government Printing Office, Washington, D.C.

_____ 1914. Abrasive Materials. In *Mineral Resources of the United States, Calendar Year 1913*, p. 253–272. Part II—Nonmetals. Department of the Interior, U.S. Bureau of Mines, Government Printing Office, Washington, D.C.

_____ 1916. Abrasive Materials. In *Mineral Resources of the United States, 1914*, p. 549–568. Part II—Nonmetals. Department of the Interior, U.S. Bureau of Mines, Government Printing Office, Washington, D.C.

_____ 1917. Abrasive Materials. In *Mineral Resources of the United States, 1915*, p. 65–80. Part II—Nonmetals. Department of the Interior, U.S. Bureau of Mines, Government Printing Office, Washington, D.C.

_____ 1919. Abrasive Materials. In *Mineral Resources of the United States, 1916*, p. 197–212. Part II—Nonmetals. Department of the Interior, U.S. Bureau of Mines, Government Printing Office, Washington, D.C.

_____ 1920. Abrasive Materials. In *Mineral Resources of the United States, 1917*, pp. 213–232. Part II—Nonmetals. Department of the Interior, U.S. Bureau of Mines, Government Printing Office, Washington, D.C.

_____ 1921. Abrasive Materials. In *Mineral Resources of the United States, Calendar Year 1918*, pp. 1171–1187. Part II—Nonmetals. Department of the Interior, U.S. Bureau of Mines, Government Printing Office, Washington, D.C.

_____ 1926. Abrasive Materials. In *Mineral Resources of the United States, 1923*, pp. 327–337. Part II—Nonmetals. Department of the Interior, U.S. Bureau of Mines, Government Printing Office, Washington, D.C.

_____ 1927. Abrasive Materials. In *Mineral Resources of the United States, 1924*, pp. 241–252. Part II—Nonmetals. Department of the Interior, U.S. Bureau of Mines, Government Printing Office, Washington, D.C.

_____ 1928. Abrasive Materials. In *Mineral Resources of the United States, 1925*, pp. 171–174. Part II—Nonmetals. Department of the Interior, U.S. Bureau of Mines, Government Printing Office, Washington, D.C.

Kelleher, Tom 1990. The Kingsbury Mill: Rare New England Survivor (Medfield, Massachusetts). *Old Mill News* 18 (4):8.

Kentucky Gazette 1790. Ad by James Wilkerson to employ hands to take his boats to New Orleans. *Kentucky Gazette*, March 29, Lexington, Kentucky.

_____ 1792a. Toliver Craig's ad for French Burr Millstones. *Kentucky Gazette*, May 5, Lexington, Kentucky.

_____ 1792b. Ad by John Moylan for men to take his boats to New Orleans. *Kentucky Gazette*, December 8, Lexington, Kentucky.

_____ 1796. Jacob and Henry Hoover's ad for French Burr Millstones. *Kentucky Gazette*, July 16, Lexington, Kentucky.

_____ 1799a. Thomas Hart's ad for French Burr Millstones. *Kentucky Gazette*, February 7, Lexington, Kentucky.

_____ 1799b. Ad for five pairs of Red River Millstones at Cleveland's Landing. *Kentucky Gazette*, June 13, Lexington, Kentucky.

_____ 1803a. Ad for James Morrison's Mill and Distillery in Lexington. *Kentucky Gazette*, April 26, Lexington, Kentucky.

_____ 1803b. Ad for D. M. Vicar's Hickman Mills seven miles from Lexington. *Kentucky Gazette*, September 6, Lexington, Kentucky.

_____ 1803c. Ad for John Roger's Grist Mills in Fayette County. *Kentucky Gazette*, September 20, Lexington, Kentucky.

_____ 1804a. Ad for John McCall's Grist Mill in Fayette County. *Kentucky Gazette*, January 31, Lexington, Kentucky.

_____ 1804b. Ad for John Roger's Grist Mills in Fayette County. *Kentucky Gazette*, March 13, Lexington, Kentucky.

_____ 1805a. Ad for William Watson's Grist Mills in Clark County. *Kentucky Gazette*, September 3, Lexington, Kentucky.

_____ 1805b. Ad for Elisha I. Winter, Jr.'s, Mills 15 miles from Lexington. *Kentucky Gazette*, November 21, Lexington, Kentucky.

_____ 1807. Ad for Thomas Hart, Jr.'s, Mills 10 miles from Lexington. *Kentucky Gazette*, July 21, Lexington, Kentucky.

_____ 1808. Ad for John McCall's Oak Ridge Mills in Fayette County. *Kentucky Gazette*, April 19, Lexington, Kentucky.

_____ 1809. Ad for Thomas Hart, Jr.'s, Mills 10 miles from Lexington. *Kentucky Gazette*, January 10, Lexington, Kentucky.

_____ 1815. John Bradford's for French Burr Millstones. *Kentucky Gazette*, January 23, 1815, Lexington, Kentucky.

_____ 1817. Ad for Benjamin Futhey's Mills 10 miles from Lexington. *Kentucky Gazette*, November 29, Lexington, Kentucky.

_____ 1819. Ad for Daniel Bradford's sale of Alluvion Mills in Lexington. *Kentucky Gazette*, March 5, Lexington, Kentucky.

_____ 1824. Ad for W. H. Tegarden's old mill on East Hickman 10 miles from Lexington by Attorney David A. Sayre. *Kentucky Gazette*, January 2, Lexington, Kentucky.

Kentucky Reporter 1818. Ad for Red River Millstones at quarry by Spencer Adams and James Daniel. *Kentucky Reporter*, April 8, Lexington, Kentucky.

Killebrew, J. B. 1874. *Introduction to the Resources of Tennessee*. First and Second Reports of the Bureau of Agriculture for the State of Tennessee. Travel, Eastman, and Howell, Nashville.

Kinnaird, Lawrence (editor) 1946. Spain in the Mississippi Valley, 1765–1794, Translations of Materials from the Spanish Archives in the Bancroft Library. Part II: Post War Decade, 1782–1791. In the *Annual Report of the American Historical Association for the Year 1945*, Volume III. U.S. Government Printing Office, Washington, D.C.

Krauss, Jewell Bell 1983. *Adams-Bell Genealogies and Allied Families*. Stevens Publishing, Astoria, Illinois.

Ladoo, Raymond B. 1925. *Non-Metallic Minerals: Occurrence-Preparation-Utilization*. McGraw-Hill, New York.

_____, and W. M. Myers 1951. *Monometallic Minerals*. Second edition. McGraw-Hill, New York.

Lawson, Rowena 1985. *Montgomery County, Kentucky 1810–1840 Censuses*. Heritage Books, Bowie, Maryland.

_____ 1986. *Montgomery County, Kentucky 1850 Census*. Heritage Books, Bowie, Maryland.

Leffler, Dankmar 2001. *Das Crawinkler Mühlsteingewerbe: Zur Geschichte eines der ältesten Gewerbe im Thüringer Wald.* Förderverein "Alte Mühle" e. V., Crawinkel. [Germany]

Lemon, J. R. 1894. *Lemon's Hand Book of Marshall County Giving Its History, Advantages, etc. and Biographical Sketches of Its Prominent Citizens.* J. R. Lemon, Benton, Kentucky.

Lexington Observer & Kentucky Reporter 1836a. Printer's ad for Millstones. *Lexington Observer & Kentucky Reporter*, June 1, p. 3, Lexington, Kentucky.

_____ 1836b. Ad for Francis F. Jackson's Grist Mill in Clark County. *Lexington Observer & Kentucky Reporter*, July 27, p. 3, Lexington, Kentucky.

Linney, W. M. 1884. *Report on the Geology of Montgomery County.* New Series, No. 2. Kentucky Geological Survey, Frankfort.

Loughridge, R. H. 1888. *Report on the Geological and Economic Features of the Jackson Purchase Region, Embracing the Counties of Ballard, Calloway, Fulton, Graves, Hickman, McCracken, and Marshall.* Geological Survey of Kentucky. Series 2, Volume F. John D. Woods, Frankfort.

Lynchburg Daily Virginian 1853. Ad by Israel Price and Company for the Brush Mountain Millstone Quarry in Montgomery County, Virginia. *Lynchburg Daily Virginian*, September 1853.

Major, J. Kenneth 1982. The Manufacture of Millstones in the Eifel Region of Germany. *Industrial Archaeology Review* 6 (3):194–204.

Mangartz, Fritz 2006. Prehistoric to Medieval Quernstone Production in the Bellerberg Volcano Lava Stream Near Mayen, Germany. *Quern Study Group Newsletter* 7:10–13. Translated by Caroline Rann. Oxford, England.

_____, and Olaf Pung 2002. Die Holzkeilspaltung im alten Steinabbau. *Der Anschnitt. Zeitschrift für Kunst und Kultur im Bergbau*, 54e année, 6/2002, pp. 238–252. [Use of wooden wedges in stone quarrying in Germany]

Manufacturer and Builder 1872. The Ventilation of Unwholesome Manufactories. *Manufacturer and Builder* 4 (4):81, New York.

_____ 1876. Burr stones. *Manufacturer and Builder* 8 (4):82, New York.

_____ 1879. Ulster County Mill Stones. *Manufacturer and Builder* 11 (10):226–227, New York.

_____ 1886. Mineral Production of the United States in 1885. *Manufacturer and Builder* 18 (11): 250–251, New York.

Martin, Larry D. 1999. *Powell County Cemeteries.* Red River Historical Society, Clay City, Kentucky.

Mather, William W. 1839. *Report on the Geological Reconnaissance of Kentucky, Made in 1838.* Journal of the Senate of the Commonwealth of Kentucky, Frankfort. Reprinted by the Kentucky Geological Survey, Series 11, Reprint 25, 1988, Lexington.

_____ 1843. *Geology of New-York.* Part I: Comprising the Geology of the First District. Carroll and Cook, Albany. Printed with permission of the New York State Museum, Albany, N.Y.

Maxwell, Hu. 1968. *The History of Barbour County, West Virginia: From Its Earliest Exploration and Settlement to the Present Time.* McClain Printing, Parsons, West Virginia.

The Maysville Eagle 1828. Ad for Morris & Egenton French Burr Millstone Manufactory of Baltimore for a branch office in Cincinnati, Ohio. *The Maysville Eagle*, January 23, Maysville, Kentucky.

_____ 1834a. Ad for millstones for sale. *The Maysville Eagle*, November 6, Maysville, Kentucky.

_____ 1834b. Ad for Raccoon Burr Millstones for sale. *The Maysville Eagle*, November 20, Maysville, Kentucky.

M'Cauley, I. H., J. L. Suesserott, and D. M. Kennedy 1878. *Historical Sketch of Franklin County, Pennsylvania: Prepared for the Centennial Celebration Held at Chambersburg, Penn'a, July 4th, 1876 and Subsequently Enlarged.* D. F. Pursel, Chambersburg, Pennsylvania.

McCalley, Henry 1886. *Geological Survey of Alabama.* Barrett, Montgomery, Alabama.

McDowell, Robert C. 1978. *Geologic Map of the Levee Quadrangle, East-Central Kentucky.* U.S. Geological Survey, Reston, Virginia.

McGee, Marty 2001. *Meadows Mills: The First Hundred Years.* Meadows Mills, Inc., Wilkesboro, North Carolina.

McGill, William M. 1936. *Outline of the Mineral Resources of Virginia.* Virginia Geological Survey, Bulletin 47, Educational Series No. 3.

McGrain, John W. 1982. "Good Bye Old Burr": The Roller Mill Revolution in Maryland, 1882. *Maryland Historical Magazine* 77 (2):154–171.

_____ 1991. Fact cards on millstones from the personal files of John W. McGrain, Baltimore County Landmarks Commission, Towson, Maryland. Photocopies of cards shared with Charles D. Hockensmith on February 4, 1991.

McIlhaney, Calvert W. [Bristol, Virginia] 1992. Personal communication to Charles D. Hockensmith. [Discussing conglomerate millstone quarries on Cloyd Mountain in Pulaski County, Virginia]

McKechinie, Jean L. (editor) 1978. *Webster's New Twentieth Century Dictionary of the English Language.* Unabridged, second edition. Collins and World.

McKee, Harley J. 1971. Early Ways of Quarrying and Working Stone in the United States. *APT* 3 (1): 44–58. Bulletin of the Association for Preservation Technology.

_____ 1973. *Introduction to Early American Masonry: Stone, Brick, Mortar and Plaster.* National Trust for Historic Preservation and Columbia University.

McMurtrie, H. 1819. *Sketches of Louisville and Its Environs.* S. Penn, Louisville, Kentucky.

Meadows, Larry G. 2002. Red River Millstone Quarry. *The Millstone* 1 (2):11–12. Kentucky Old Mill Association, Clay City, Kentucky.

_____ 2006. Mills, Stills and Other Deals. *The Millstone* 5 (2):37–41. Kentucky Old Mill Association, Clay City, Kentucky.

Metcalf, Robert W. 1941. Abrasive Materials. In *Minerals Yearbook Review of 1940*, pp. 1239–1254. Department of the Interior, U.S. Bureau of Mines, Government Printing Office, Washington, D.C.

_____ 1943a. Abrasive Materials. In *Minerals Yearbook, 1941*, pp. 1339–1356. Department of the Interior, U.S. Bureau of Mines, Government Printing Office, Washington, D.C.

_____. 1943b. Abrasive Materials. In *Minerals Yearbook, 1942*, pp. 1331–1348. Department of the Interior, U.S. Bureau of Mines, Government Printing Office, Washington, D.C.

_____. 1949. Abrasive Materials. In *Minerals Yearbook, 1947*, pp. 97–113. Department of the Interior, U.S. Bureau of Mines, Government Printing Office, Washington, D.C.

_____ 1950. Abrasive Materials. In *Minerals Yearbook, 1948*, pp. 98–115. Department of the Interior, U.S. Bureau of Mines, Government Printing Office, Washington, D.C.

_____. 1951. Abrasive Materials. In *Minerals Yearbook, 1949*, pp. 91–110. Department of the Interior, U.S. Bureau of Mines, Government Printing Office, Washington, D.C.

_____, and A. B. Cade 1945. Abrasive Materials. In *Minerals Yearbook, 1943*, pp. 1384–1399. Department of the Interior, U.S. Bureau of Mines, Government Printing Office, Washington, D.C.

_____, and _____ 1946. Abrasive Materials. In *Minerals Yearbook, 1944*, pp. 1341–1358. Department of the Interior, U.S. Bureau of Mines, Government Printing Office, Washington, D.C.

_____, and A. B. Holleman 1947. Abrasive Materials. In *Minerals Yearbook, 1945*, pp. 1357–1376. Department of the Interior, U.S. Bureau of Mines, Government Printing Office, Washington, D.C.

_____, and _____ 1948. Abrasive Materials. In *Minerals Yearbook, 1946*, pp. 92–110. Department of the Interior, U.S. Bureau of Mines, Government Printing Office, Washington, D.C.

Michael, Ronald L. 1983. National Register of Historic Places Nomination Form for Game Lands 51 Millstone Quarry, Fayette County, Pennsylvania. Prepared by California University, California, Pennsylvania.

Michaux, F. A. 1805. Travels to the West of the Alleghany Mountains in the States of Ohio, Kentucky, and Tennessea and Back to Charleston, by the Upper Carolines. Reprinted in *Early Western Travels 1748–1846*, edited by Reuben Gold Thwaites (1904), Volume 3, pp. 105–306. Arthur H. Clark, Cleveland.

Miller, Mrs. Ralph G. 1971. Spencer Adams (175?-1830) of Virginia, North Carolina, Kentucky, and Alabama: Some of His Brothers, Some of His Descendants, and a Few of His Associates. Manuscript, 21 pp. Kentucky Historical Society Library, Frankfort.

Montgomery County 1907. *Montgomery County, Virginia*. Jamestown Exposition Souvenir. Blacksburg, Virginia.

Moore, Lucas 1897. *Twelfth Biennial Report of the Bureau of Agriculture, Labor and Statistics of the State of Kentucky*. Geo. G. Fetter Printing, Louisville, Kentucky.

The Morning Star 1887. Comment concerning the Moore County, North Carolina, millstone industry. *The Morning Star*, April 29, Wilmington, North Carolina.

Morton, Lynn Douglas 1994. *1880 Census, Powell County, Kentucky*. Privately printed, Stanton, Kentucky.

_____ 1995. *1900 Census, Powell County, Kentucky*. Privately printed, Stanton, Kentucky.

Morton, Mildred Napier *1920. Census of Powell County, Kentucky*. Privately printed, Stanton, Kentucky.

Nason, F. L. 1894. *Economic Geology of Ulster County*. New York State Museum 47th Annual Report, Albany.

New York State Museum 1918. Millstone Producers in New York—1918, Ulster County. Typed list on file with the Geological Survey, New York State Museum, Albany. Printed with permission of the New York State Museum, Albany, N.Y.

_____ 1934. Millstones in 1934, New York, Ulster County. Typed list on file with the Geological Survey, New York State Museum, Albany. Printed with permission of the New York State Museum, Albany, N.Y.

Newland, David H. 1907. Report of Operations and Productions during 1906. In *The Mining and Quarry Industry of New York State 1906*. New York State Museum Bulletin Number 305, Albany.

_____ 1908. *The Mining and Quarry Industry of New York State 1907*. New York State Museum Bulletin Number 426, Albany.

_____ 1909. *The Mining and Quarry Industry of New York State 1908*. New York State Museum Bulletin Number 451, Albany. Printed with permission of the New York State Museum, Albany, N.Y.

_____ 1910. *The Mining and Quarry Industry of New York State 1909*. New York State Museum Bulletin Number 476, Albany. Printed with permission of the New York State Museum, Albany, N.Y.

_____ 1911. *The Mining and Quarry Industry of New York State 1910*. New York State Museum Bulletin Number 496, Albany. Printed with permission of the New York State Museum, Albany, N.Y.

_____ 1916. *The Mining and Quarry Industry of New York State 1915*. New York State Museum Bulletin Number 190, Albany. Printed with permission of the New York State Museum, Albany, N.Y.

_____ 1921. *Mineral Resources of the State of New York for 1919*. New York State Museum Bulletins, Numbers 223 and 224, Albany. Printed with permission of the New York State Museum, Albany, N.Y.

_____, and C. A. Hartnagel 1932. *Mining and Quarry Industries of New York 1927–1929*. New York State Museum Bulletin Number 295, Albany. Printed with permission of the New York State Museum, Albany, N.Y.

_____, and _____ Hartnagel 1936. *Mining and Quarry Industries of New York 1930–1933*. New York State Museum Bulletin Number 305, Albany. Printed with permission of the New York State Museum, Albany, N.Y.

_____, and _____ Hartnagel 1939. *Mining and Quarry Industries of New York 1934–1936*. New York State Museum Bulletin Number 319, Albany. Printed with permission of the New York State Museum, Albany, N.Y.

Norris, William V. 1981. *1860 U.S. Census, Clark County, Kentucky and Index*. Privately printed, Jacksonville, Florida.

_____. 1983. *1840 U.S. Census, Clark County, Kentucky and Index*. Privately printed, Jacksonville, Florida.

North Carolina Land Company 1869. *A Statistical and Descriptive Account of the Several Counties of North Carolina, United States of America*. Nichols & Gorman, Raleigh.

North Carolina State Geologist 1875. *Report of the Geological Survey of North Carolina*. J. Turner, State Printer, Raleigh.

Observer & Reporter 1839. I. I. McConathy's ad for millstones. *Observer & Reporter*, January 2, page 1, Lexington, Kentucky.

Owen, David Dale 1858. *First Report of a Geological Reconnaissance of the Northern Counties of Arkansas, Made during the Years 1857 and 1858*. Arkansas Geological Survey. Johnson & Yerkes, Little Rock.

_____ 1861. *Fourth Report of the Geological Survey in Kentucky Made during the Years 1858 and 1859*. Printed at the Yeoman Office, Frankfort.

Owen, Katheryn 1975. *Old Graveyards of Clark County, Kentucky*. Polyanthos, Inc., New Orleans.

_____, and Anne P. Couey 1983. *Early Winchester Cemetery Inscriptions, Winchester, Clark County, Kentucky*. McDowell Publications, Utica, Kentucky.

The Palladium 1800. Ad for John Tanner's millstone quarry in Woodford County, Kentucky. *The Palladium*, February 27, Frankfort, Kentucky.

Paris Western Citizen 1817. Ad for sale of Alexander Ogle Grist and Saw Mill. *Paris Western Citizen*, August 5, 1817, Paris, Kentucky.

Parker, Edward W. 1893a. Abrasive Materials. In *Mineral Resources of the United States, Calendar Year 1891*, pp. 552–554. Part II — Nonmetals. U.S. Bureau of Mines, U.S. Geological Survey, Government Printing Office, Washington, D.C.

_____ 1893b. Abrasive Materials. In *Mineral Resources of the United States, Calendar Year 1892*, pp. 748–750. Part II — Nonmetals. U.S. Bureau of Mines, U.S. Geological Survey, Government Printing Office, Washington, D.C.

_____ 1894. Abrasive Materials. In *Mineral Resources of the United States, Calendar Year 1893*, pp. 670–679. Part II — Nonmetals. U.S. Bureau of Mines, U.S. Geological Survey, Government Printing Office, Washington, D.C.

_____ 1895. Abrasive Materials. In *Sixteenth Annual Report of the United States Geological Survey to the Secretary of the Interior 1894–1895*, by Charles D. Walcott, pp. 927–950. Part III — Mineral Resources of the United States Geological Survey, Government Printing Office, Washington, D.C.

_____ 1896. Abrasive Materials. In *Seventeenth Annual Report of the United States Geological Survey to the Secretary of the Interior 1895–1896*, by Charles D. Walcott, pp. 586–587. Part IV — Mineral Resources of the United States. Washington, D.C.

_____ 1897. Abrasive Materials. In *Eighteenth Annual Report of the United States Geological Survey to the Secretary of the Interior 1896–1897*, by Charles D. Walcott, pp. 1219–1231. Part V — Mineral Resources of the United States. Washington, D.C.

_____ 1898. Abrasive Materials. In *Nineteenth Annual Report of the United States Geological Survey to the Secretary of the Interior 1897–1898*, by Charles D. Walcott, pp. 515–533. Part VI — Mineral Resources of the United States. Washington, D.C.

Patrick, Tracy R. 1981. *The 1860 Census of Powell County, Kentucky*. Kinko's Copies, Richmond, Kentucky.

_____ 1988. *1870 Census, Powell County, Kentucky*. The Estill Tribune, Irvine, Kentucky.

Peters, J. T., and H. B. Carden 1926. *History of Fayette County, West Virginia*. Jarrett Printing Company, Charleston, West Virginia.

Phalen, W. C. 1908. Abrasive Materials. In *Mineral Resources of the United States, Calendar Year 1907*, pp. 607–626. Part II — Nonmetals. Department of the Interior, U.S. Bureau of Mines, Government Printing Office, Washington, D.C.

_____ 1909. Abrasive Materials. In *Mineral Resources of the United States, Calendar Year 1908*, pp. 581–598. Part II — Nonmetals. Department of the Interior, U.S. Bureau of Mines, Government Printing Office, Washington, D.C.

_____ 1910. Abrasive Materials. In *Mineral Resources of the United States, Calendar Year 1909*, pp. 609–627. Part II — Nonmetals. Department of the Interior, U.S. Bureau of Mines, Government Printing Office, Washington, D.C.

_____ 1911. Abrasive Materials. In *Mineral Resources of the United States, Calendar Year 1910*, pp. 683–690. Part II — Nonmetals. Department of the Interior, U.S. Bureau of Mines, Government Printing Office, Washington, D.C.

_____ 1912. Abrasive Materials. In *Mineral Resources of the United States, Calendar Year 1911*, pp. 835–854. Part II — Nonmetals. Department of the Interior, U.S. Bureau of Mines, Government Printing Office, Washington, D.C.

Pigg Family 1985. *A Pigg Family Newsletter* 1 (1): Possum Trot University Press, Manchester, Kentucky.

Pratt, Joseph H. 1901. Abrasive Materials. In *Mineral Resources of the United States, Calendar Year 1900*, pp. 787–801. Part II — Nonmetals. Department of the Interior, U.S. Bureau of Mines, Government Printing Office, Washington, D.C.

_____ 1902. Abrasive Materials. In *Mineral Resources of the United States, Calendar Year 1901*, pp. 781–809. Part II — Nonmetals. Department of the Interior, U.S. Bureau of Mines, Government Printing Office, Washington, D.C.

_____ 1904a. Abrasive Materials. In *Mineral Resources of the United States, Calendar Year 1902*, pp. 873–890. Part II — Nonmetals. Department of the Interior, U.S. Bureau of Mines, Government Printing Office, Washington, D.C.

_____ 1904b. Abrasive Materials. In *Mineral Resources of the United States, Calendar Year 1903*, pp. 989–1015. Part II — Nonmetals. Department of the Interior, U.S. Bureau of Mines, Government Printing Office, Washington, D.C.

_____ 1905. Abrasive Materials. In *Mineral Resources of the United States, Calendar Year 1904*, pp. 995–1015. Part II — Nonmetals. Department of the Interior, U.S. Bureau of Mines, Government Printing Office, Washington, D.C.

_____ 1906. Abrasive Materials. In *Mineral Resources of the United States, Calendar Year 1905*, pp. 1069–1085. Part II — Nonmetals. Department of the Interior, U.S. Bureau of Mines, Government Printing Office, Washington, D.C.

The Public Advisor 1827. Ad for sale of French Burr Millstones in Louisville. *The Public Advisor*, May 15, Louisville, Kentucky.

Radley, Jeffrey 1966. Peak Millstones and Hallamshire Grindstones. *The Newcomen Society for the Study of the History of Engineering and Technology Transactions* 36:165–173, London. [England]

Rawson, Marion N. 1935. *Little Old Mills*. E. P. Dutton, New York.

Reinemund, John A. 1955. *Geology of the Deep River Coal Field, North Carolina*. Geological Survey Professional Paper 246. U.S. Geological Survey, Washington, D.C.

Rice, C. L., and G. W. Weir 1984. Lee and Breathitt Formations along the Northwestern Part of the Eastern Coal Field. In *Sandstone Units of the Lee Formation and Related Strata in Eastern Kentucky*. Geological Survey Professional Paper 1151-6. U.S. Government Printing Office, Washington, D.C.

Ridenour, George L. 1977. *Early Times in Meade County, Kentucky*. Ancestral Trails Historical Society, Vine Grove, Kentucky. Reprint of 1929 edition published by Western Recorder, Louisville, Kentucky.

Robertiello, Barbara 1994. Vincient Dunn Remembers a Pataukunk Boyhood and More. *The Accordian* 8 (4):1–7, Rochester, New York.

_____ 1995. An Odyssey, a Dream, and a Machine Shop: Otto Paul Tolski Tells His Story. *The Accordian* 9 (3):7–10, Rochester, New York.

Roberts, Ellwood 1904. Biographical Annuals of Montgomery County, Pennsylvania: Containing Genealogical Records of Representative Families, Including Many Early Settlers and Biographical Sketches of Prominent Citizens. T. S. Benham, New York.

Rockwell, Peter 1993. *The Art of Stoneworking: A Reference Guide*. Cambridge University Press, New York.

Rogers, Diane 1996. *Court Records — Volume VII, Powell County, Kentucky*. The Red River Historical Society, Clay City, Kentucky.

_____ 1998. *Court Records — Volume XIII, Powell County, Kentucky*. The Red River Historical Society, Clay City, Kentucky.

Rogers, Lillian Berry 1986. *The Berry's and Related Families 1650 to 1986*. Privately printed, Madison Heights, Michigan.

Rogers, William Barton 1884. *A Reprint of the Annual Reports and Other Papers, on the Geology of the Virginias*. Original papers from 1835 to 1841. D. Appleton, New York.

Rotenizer, David E. [Blacksburg, Virginia] 1989. Letter to Charles D. Hockensmith. Copy on file at the Kentucky Heritage Council, Frankfort. [Discussing the Brush Mountain millstone quarry, Montgomery County, Virginia]

Russell, John 1949. Millstones in Wind and Water Mills. *The Newcomen Society for the Study of the History of Engineering and Technology Transactions* 24:55–64, London.

Safford, James M. 1869. *Geology of Tennessee*. S. C. Mercer, Nashville.

Sass, Jon A. 1984. *The Versatile Millstone: Workhorse of Many Industries*. Society for the Preservation of Old Mills, Knoxville.

Schrader, Frank C., Ralph W. Stone, and Samuel Sanford 1917. *Useful Minerals of the United States*. Bulletin 624. U.S. Geological Survey, Washington, D.C.

Scientific American 1852. Ad for French Burr Stones by Munn & Company. *Scientific American*, Volume 7, Issue 18, January 17, page 143, New York.

_____ 1863. Ad for four pairs of French Burr Stones. *Scientific American*, Volume 8, Issue 15, April 11, page 238, New York.

Shane, John D. n.d. Interview with William Risk. 11CC86, Draper Collection (Kentucky Papers). Microfilm copy at the Kentucky Historical Society, Frankfort.

Sopko, Joseph 1991. Memorandum to Charles Florance, January 28, 1991. New York State Office of Parks, Recreation, and Historic Preservation, Waterford, New York. Copy on file at Kentucky Heritage Council, Frankfort. [Discussing Ulster County millstone industry]

Spafford, Horatio Gates 1813. *A Gazetteer of the State*

of New-York; Published by H. C. Southwick, Albany.

Sterrett, Douglas B. 1907. Abrasive Materials. In *Mineral Resources of the United States, Calendar Year 1906*, pp. 1043–1054. Part II — Nonmetals. Department of the Interior, U.S. Bureau of Mines, Government Printing Office, Washington, D.C.

Stoner, Jacob 1947. Old Millstones, May 31, 1934. In *Historical Papers, Franklin County and the Cumberland Valley, Pennsylvania* by Jacob H. Stoner and Lu Cole Stoner, pp. 411–430. Craft Press, Chambersburg, Pennsylvania.

Storck, John, and Walter D. Teague 1952. *Flour for Man's Bread: A History of Milling*. University of Minnesota Press, Minneapolis.

Strawhacker, William 2004. E-mails to Charles D. Hockensmith, January 19, January 30, February 7. [Discussing the involvement of the Strohecker family of Berks County, Pennsylvania, in millstone making]

Sullivan, George M. 1891. *Report on the Geology of Parts of Jackson and Rockcastle Counties*. Kentucky Geological Survey, Series 2, 20 pages, Frankfort.

Sussenbach, Tom 1990. Personal communication to Charles D. Hockensmith, December 7. [Discussing a report of a millstone quarry on Indian Creek in McCreary County, Kentucky]

_____ 1991. Personal communication to Charles D. Hockensmith. [Discussing a report of a millstone quarry near Cumberland Falls in Whitley County, Kentucky]

Swift, Michael 1988. The Millstone Industry. *The Accordian* 2 (2):6–7, Rochester, New York.

Swisher, Jacob A. 1940. *Iowa, Land of Many Mills*. The State Historical Society of Iowa, Iowa City.

Talbert, Charles G. 1962. *Benjamin Logan, Kentucky Frontiersman*. University of Kentucky Press, Lexington.

T. L. C. Genealogy 1990. *Clark County, Kentucky Taxpayers, 1793 Thru 1799*. T. L. C. Genealogy, Miami Beach, Florida.

Todd, Levi, et al. 1965. *Some Pre-1800 Kentucky Tax Lists for the Counties of: Fayette—1788, Mason (Later Floyd) 1790, Mercer—1789, Washington—1792*. Borderland Books, Anchorage, Kentucky.

Tomlinson, Tom D. 1981. *Querns, Millstones, and Grindstones Made in Hathersage & District*. Hathersage Parochial Church Council, Sheffield. 20 pages. [England]

Treadway, William E. 1951. *Treadway and Burket Families: A Merger of the Genealogical Histories of the Treadway and Burket Families in America, Through the Documented Ancestry of Jonas Robert Treadway*. Privately printed, Topeka, Kansas.

Truax, J.W. 1896. *The Eagle Mill Pick*. J.W. Truax Firm, Essex Junction, Vermont.

Tucker, D. Gordon 1977. Millstones, Quarries and Millstone Makers. *Post-Medieval Archaeology* 11:1–21. [England]

_____ 1980. Millstone Making in Anglesey. *Wind and Water Mills* 1:16–23 [England]

_____ 1982. Millstone Making in France: When Éperson Produced Millstones. *Wind and Water Mills* 3, 32 pages. [Translation of article published in French]

_____ 1985. Millstone Making in the Peak District of Derbyshire: The Quarries and the Technology. *Industrial Archaeology Review* 8 (1):42–58. [England]

Tuomey, M. 1848. *Report on the Geology of South Carolina*. A. S. Johnston, Columbia, South Carolina.

U.S. Secretary of State 1823. *Digest of Accounts of Manufacturing Establishments in the United States and of Their Manufactures*. Made under direction of the Secretary of State, in pursuance of a Resolution of Congress, of 30th March, 1822. Printed by Gales & Seaton, Washington, D.C.

Verhoeff, Mary 1917. *The Kentucky River Navigation*. Filson Club Publication No. 28. John P. Morton and Company, Louisville, Kentucky.

Wallcut, Thomas 1879. *Journal of Thomas Wallcut, in 1790*. University Press. J. Wilson and Son, Cambridge, Massachusetts.

Ward, Owen H. 1982a. Millstones from La Ferté-sous-Jouarre, France. *Industrial Archaeology Review* 6 (3):205–210. [France]

_____ 1982b. French Millstones. *Wind and Water Mills* 3:36–43. [France]

_____ 1984a. The Making and Dressing of French-Burr Millstones in France in 1903. *Wind and Water Mills* 5:27–32. [France]

_____ 1984b. The Slaughter of the French Millstonemakers. *BIAS Journal* 17:30–31. [France]

_____ [Bath, England] 1994. Letter to Charles D. Hockensmith, December 24.

Watson, Thomas 1907. *Mineral Resources of Virginia*. J. P. Bell, Lynchburg, Virginia.

Weaks, Mabel Clare 1925. *Calendar of Kentucky Papers of the Draper Collection of Manuscripts*. Publication of the Historical Society of Wisconsin, Calendar Series II, Madison.

Weaver, Valerie 1995. The Lawrence Brothers Share 80+ Years of Saint Josen Memories. *The Accordian* 9 (2):1–5, Rochester, New York.

Webb, William S. 1933. The Millstone as an Antique. *Kentucky School Journal*. March, pp. 30–34.

_____ 1935. Old Millstones of Kentucky. *The Filson Club History Quarterly* 9 (4):209–221, Louisville, Kentucky.

Wieck, Dorothy L. 1983. *1860 Census Index for Kenton County, Kentucky*. Kenton County Historical Society, Covington, Kentucky.

_____ 1986. *1870 Census Index for Kenton County, Kentucky*. Kenton County Historical Society, Covington, Kentucky.

_____ 1996. *Kenton County, Ky. Census of 1880*. 2 volumes. Kenton County Historical Society, Covington, Kentucky.

Wilkie, Aitken 1874. The Only True and Practical

Way to Make Edge Tools: Adapted for Edge Tool-Makers, Millers for Mill-Picks, Blacksmiths, and all Kinds of Tool-Dressers and Steelworkers. Blade Printing and Paper Company, Toledo, Ohio.

Williams, Albert, Jr. 1885. Abrasive Materials. In *Mineral Resources of the United States, Calendar Years 1883–1884,* pp. 581–594. Part II — Nonmetals. U.S. Bureau of Mines, U.S. Geological Survey, Government Printing Office, Washington, D.C.

Williams & Company 1861. *Williams' Covington Directory*. Williams & Company, Cincinnati, Ohio.

Wonn, Mildred 1981. *The 1860 Census of Powell County, Kentucky*. Privately printed, Owingsville, Kentucky.

Worsham, Gibson 1986a. *Montgomery County Historic Sites Survey, Volume 1*. Unpublished report submitted to the Montgomery County Planning Commission and the Virginia Division of Historic Landmarks, Richmond.

_____ 1986b. *Montgomery County Historic Sites Survey, Volume 2*. Unpublished report submitted to the Montgomery County Planning Commission and the Virginia Division of Historic Landmarks, Richmond.

Zerfass, Samuel G. 1921. Souvenir Book of the Ephrata Cloister: Complete History from Its Settlement in 1728 to the Present Time: Included Is the Organization of Ephrata Borough and Other Information of Ephrata Connected with the Cloister. John G. Zook, Lititz, Pennsylvania.

Index

Accord, New York xii, 139
The Accordian xii, xiii
Adams, Elkanah 32, 33, 39, 40
Adams, Gary xiii
Adams, Harrison 40
Adams, John 30
Adams, John A. 30
Adams, John A., Jr. 32
Adams, Joseph 12
Adams, Margery 32
Adams, Mary Ann (Pigg) 39
Adams, Nathan 28, 30–31
Adams, Patsy 31
Adams, Payton 31
Adams, Peyton 31
Adams, Polly Eubank 31
Adams, R. Foster 181
Adams, Robert 31–32
Adams, Rose 31
Adams, Sara Corbin 40
Adams, Sara Pigg 32
Adams, Sarah M. 31
Adams, Sonella Jane 28
Adams, Spence 32
Adams, Spencer x, 15, 26, 27, 32–33, 34, 36, 39, 40, 41, 43, 159, 173–178
Adams, William G. x, 181
Alexander Hamilton Memorial Free Library xii
Allan & Simpson 41, 175
Allegheny Mountains 164
Allen, Joseph 13
Allins, Dillard C. 172
Alluvion Mills 150, 155
Alvis, Gayle x
Ambrose, Paul M. 181
American beech 21
American chestnut 21
American Geological Institute xiii
American hazelnut 21
American Miller 181
Anderson, William H. 41
Anderson County, Kentucky 38
Anglesey 139
Apperson, Richard 181
apple cider 8
archaeological investigations 47–111
Archaeological Society of Virginia x
archaeologists ix, x, 9, 160, 161
Archival Research 24–25
The Argus of Western America 11, 15, 16, 25, 157, 194
Arlington, Virginia x

Arndt, Karl J. R. 181
Arnow, Harriette S. 181
artifacts x, xiv, 123–127
Athens County, Ohio 165

Baber, Adin 181
Bailey, Harry H. 181
Baird & Owen's Store 152
Baker, Darlene H. 29
Baker, Donald 29
Baker, Ira O. 181
Baker, James xi, 29
Baker, John L. 13
Baker, Mary Jo 29
Baker, Ronnie xi, 29
Baker Millstone Quarry x, xi, xii, 29, 63–76, 112–115, 117–123, 125–133, 179
Ball, Donald B. x, 181
Baltimore American 182
Baltimore County, Maryland 44
Baltzell, George 12
Baltzell, John 12
Barnes & Pearsol 182
Barren River 149
barytes 8
basalt 8
basswood 201
Beach, L. M. 182
bedstone 7, 16, 159, 163, 172–173
beech 20
Beech Fork 40
Belmont, Alain 138, 182
benches 49, 78, 104
Bennette, Nancy 28
Bennette, Thomas 28
Berg, Thomas 182
Berry, Abigail 28, 33
Berry, Albie 33
Berry, Benjamin 33
Berry, Dora 29
Berry, Elizabeth 33
Berry, I. N. 28
Berry, Isaac Newton 33–34
Berry, Margaret Newton 35
Berry, Newton 28
Berry, Peter DeWitt 28, 33, 34
Berry, Sally DeWitt 34
Berry, Sarah DeWitt 28, 33, 34
Berry, Thomas 33, 34, 35
Berry, Thomas, Jr. 34
Berry, Thomas, Sr. 34
Berry, Thomas H. 34
Big Barren River 13
bitternut hickory 20

Black Creek 15, 26, 27, 28, 29, 30, 33, 38
black gum 21
black jack oak 21
black locust 21
black oak 21
black walnut 21
Blacksburg, Virginia xi, 141
Bledsoe Creek 13, 154
blocking hammer 137, 163
blueberry 21
Bluegrass Region 8, 19, 148, 150
Bourbon County, Kentucky 16, 30, 33, 42, 109, 153, 155, 178, 180
boulder and drill hole recording form 170–172
Bowles, Oliver 182
Bowling Green, Kentucky x
Bowman, James F. 182
Boyd, Hazel Mason 182
Bradford, Daniel 150, 155
Bradford, James 155
Bradford, John 153
Bradford, T., and Company 155, 156
Braun, E. Lucy 182
Bray, Jane 36
bread 1
Briggs, Joe ix
Briggs, Guy H., Jr. xiv, 9
Brink, John A. 46
Bristol, Virginia x
Brose, David S. 183
Broughton, John G. 185
Brown, Bruce B. xi, 182
Brown, Richard xii
Brundage, Larry 182
Brush Creek 15, 27, 28, 29, 33, 35, 36, 179
Brush Mountain Millstone Quarry xi, 136, 163
Bryan, David xiv
Bryant, Ron x
Buchanan & Starkey 153
buckeye 20
Buckley, Jeremiah 11, 12, 157
Buckley's Ferry 12
Buford, John 12
buhrstone (burr stone, etc.) 17, 18, 163
bull point 139
bull riggings 137
Bullock, James M. 181
bullrigging 163
bullset 137, 142, 146, 163
bush hammer 137, 163
Butler, Brian M. 182

197

Index

Cade, A. B. 191
Caldwell County, Kentucky 12
calipers 163
Calk, William 174
Call, John M. 154
Calmes, Marquis, Jr. 33
Calmes, Marquis, Sr. 33
Campbell, Marius R. 182
Carden, H. B. 192
cement 8
Cesari, Bernard xiii
chaffing wheat 8
Chandler, Henry P. 182, 183
channel fills 19–20
charcoal 8
chasers 8, 163
chestnut oak 20, 21
Chestnut Ridge, Pennsylvania 164
Child, Hamilton 183
chipping hammer 137, 163
chisels 137, 140, 142–143
chocolate industry 8
Christian, Patricia xii
Cinadr, Thomas J. 183
Cincinnati, Ohio 153, 155
Civil War 26, 160
Clark, Jas. E. 177
Clark, Victor S. 183
Clark County, Kentucky xii, 14, 15, 25, 26, 30, 31, 32, 33, 35, 36, 37, 38, 39, 40, 41, 42, 43, 46, 154, 159, 172–178, 185
Clark Pilot knob 15
Clay, Brutus 158
Clay, R. Berle x, 183
Clay, Sidney Payne xiii, 16
Clay City, Kentucky 149
Clay Lick Branch 30, 31
Clear Creek 12
Clemons, James Marvin 29
Clemons, Wanda M. 29
Cleveland's Landing 15, 149, 157
Clifford, J. D. 183
Clift, G. Glenn 183
Clouse, Jerry A. xi, 183
clover seeds 8
Cocalico millstones xii, 139–140, 163
Collins, Dale x
Collins, Dillard 172
Collins, Gabriel 183
Collins, Lewis 183
Collins, Richard H. 183
color mills 8
Colyer, Charles, Jr. xiii, 15, 16, 183
Colyer, Rockcastle, and Otway series soils 20
Combs, Benjamin 33
Combs, Cuthbert 33
The Commonwealth 25, 153, 194
compasses 137, 140, 141, 146
composite millstones 7, 163
conglomerate 8, 14, 15, 17, 18, 19, 51, 66–67, 78, 85, 98, 106, 112, 152, 163, 165
Conkwright, Isaac 178
Conley, James F. 183
Connell, James 46
Coons, A. T. 182, 183
Coons, John 46
Cooper, James D. 184

Corbin, Rawley 40
Corbin, Sara 32
cork mills 8
corn 8, 13, 152, 156, 159, 173
Cornell University 25
Cornett, Dover xi
cottonwood 21
Couey, Anne P. 184
Covington, Kentucky 155
Cowbell Member of the Borden Formation 19
Coy, Fred E., Jr. x, xi, xii, 187
Craig, Toliver 154
Craik, David 184
Crawford, Byron 184
Crawwinkle, Germany 138
Croft Mills 13, 154
Cross, Alice xiii, 184
cross grain 164
crow bars 137
Cross, Alice xiii, 184
cucumber tree 20
Cultural Resource Analysis, Inc. x
Cumberland Falls 17
Cumberland River 13, 17
Cuming, F. 184
cutting eyes 143, 147
cutting hammer 137, 142, 164

damsel 7, 155
Daniel, Beverly 35
Daniel, James 15, 27, 33, 35, 158, 159
Daniel, James, Jr. 35
Daniel, James, Sr. 35
Daniel, James M. 35, 36
Daniel, Mary 36
Daniel, Sarah 35
Danley, Benjamin 46
Darton, N. H. 184
Davis, A. E. 182, 184, 185, 188
Davis, L-yleton 33
Davis, Singleton 27
Davy's Fork of Elkhorn Creek 154
Day, David T. 184
Deacon, John 46
Deacon, Joseph 46
Dean, Lewis S. xi, 184
Dedrick, B. W. 184
deeds 25, 26, 27, 28, 29, 32
Dennen, Mark xiv
Derbyshire, England 165
derrick 164
Dever, Garland R. xii
dew berries 21
DeWitt, Catherine 43
DeWitt, Martin 15, 35, 36
DeWitt, Mary Catherine 44
DeWitt, Massey 36
DeWitt, Peter 15, 27, 33, 35, 36, 37, 40, 41, 44
DeWitt, Peter, Jr. 36
DeWitt, Peter, Sr. 27, 36, 159
DeWitt, Sally 35
DeWitt, Thomas 37
DeWitt, William 37
Dexter, William R. xiv
dogwood 20
Dolly, Owen 27
Doncaster, Richard 137
Drake, Ed x
drill holes 60–62, 74–75, 83–84, 96–98, 102–104, 110–111, 119–120, 121, 164
drilled boulders 7, 60–63, 74–76, 83–84, 95, 102–104, 110–111, 119–120
drills 136–137, 139, 164
drugs 8
Dudley 11
Dunn, Thelma M. 184
Durlach, Pennsylvania xii
Dyche, Russell 184
dye mills 8

The Eagle 153, 156, 194
East Hickman Creek 150, 154, 155
Ebert, Milford 184
edge runners 8, 164
Éditions Ibis Press xiii
Edwards, Richard 184
Eifel Region, Germany 138
elderberry 21
Elkhorn Creek 17
Ellenville, New York xiii
Ellenville Public Library and Museum xiii
Emmons, Ebenezer 184
England 5, 134–135, 138
English millstones 156
Enoch, Harry xiii, 184
Epernon, France 138
Ephrata, Pennsylvania xiii
Erb, Edward E. xiii
Esopus millstones 139, 164
Estill County, Kentucky 32, 38, 40, 41, 44
Eubank, Achilles S. 32
Euins, Margaret 44
Europe 7, 9, 138–139
Evans, Margaret 44
Ewen, Mrs. A. L. xi
Ewen, Sadie 34
Ewen Millstone Quarry xi, 98–104, 112–115, 117–121, 123, 127–131, 133
eyes 16, 51–52, 67–68, 80, 83, 87, 100–102, 109, 117–118, 139, 164

face grinders 8, 164
face of millstones 164
facing hammers 137, 140
Faulkner, Johnny x
Fayette County, Kentucky 12, 15, 30, 32, 33, 34, 36, 38, 42, 43, 154, 172–173, 184
Fayette County, Pennsylvania xiii, 37, 164
feathers 124, 126
Fedders, James M. 184
feldspar 8
Felldin, James Robey 184
fertilizers 8
Field Limitations 23–24
Field Methods 22–23
The Filson Historical Society xiii, 16
flat boats 149
flax mills 8
Flick, Alexander C. 185
flint 8, 11, 17, 18, 156, 158, 165
Flint, James 155, 158
flint grinding 8
Flint Ridge millstones 164

Index

Flory, Paul B. xiii, 136, 140, 185
flour 1, 7
Flourney, Matthew 12
Flowerdew Hundred 140
flowering dogwood 21
Floyd, Virginia x
Flynn, Cora Ware 29
Flynn, Samuel F. 29
Fontaine, William M. 185
Four Mile Creek 155
fragrant sumac 21
France 138, 155
Frankfort, Kentucky xiv, 3, 150, 153, 172, 173, 175, 177
The Frankfort Commonwealth 19, 155
Franklin County, Kentucky 3, 11, 150, 157
Frazier, Andy xi
French, James 127, 173
French and Farrow 178
French Burr millstones 12, 17, 138, 152–156, 164
fresh water quartz 8
Friends of Historic Rochester, Inc. xiii
Fries, Robert 185
Fuller, Tom x
furrows 7, 164
Futhey, Benjamin 150, 154

Gafney, Dale ix, x
Gallatin County, Kentucky 153
Gannett, Henry 185
Garber, D. W. 185
Gardner, Charles H. xii
Gary, Margaret 185
Gates, Janet xii
Gembe, Ruth Baer xii
Geological Survey of Alabama xi
geology 19–20
Germany 138
Gettel, Phil x
Gildersleeve, Benjamin x
Gilpin, Joshua 185
glucose 8
gneiss 8
Golf, John Hodges 37
Golf, Nancy 37
Golf, Patsy 37
Goose Creek 13, 164
Goose Creek Burrs 13, 154
Goose Creek millstones 13
Gordon, Kentucky 12
Gordon, William 158
granite 8, 165
Grape Knob 20
Grassi, Robert 185
Great Britain 25
Green County, Kentucky 38
Greencastle, Pennsylvania x, xi
Griffin, Elizabeth 43
Grimshaw, Robert 185
grinding bone 8
grindstones 15, 18, 35, 36
gun powder mills 8
gypsum grinding 8

Hadden, Elizabeth 43
Hadden, Nicholas 28
Hagley Museum and Library xii

Hailstone Grit 15, 17
Halcomb, Clarence xi, 12, 17, 185
Hamilton, Ohio xi
Hampton, Esther 35
Hampton heirs 29
Hanks, Absolom 15, 38, 150, 159, 172–173
Hanks, Absalom, Sr. 38
Hanks, Peter III 38
Hannan, Richard xi
Hardwick's Creek 149
Harms, Eduard 138, 185
Hart, Thomas, Jr. 150, 152, 154, 185
Hartnagel, C. A. 185, 192
Harvey, H. H. 185
Hasley, Verlin ix, x
Hatmaker, Paul 185
hawthorn 21
Hazen, Theodore R. 186
Hearice, Will 16
Hedger, Benjamin 27, 38
Hehnly & Wike xii, 186
Heinemann, Charles B. 186
Helton, Walter L. 186
hemp mills 8, 13
Henry County, Kentucky 39
Hensley, Mrs. Chuckie Hall 186
Herbert & Wright 155–156
Hercules' Club 21
HeritageQuest 25
Herzel, Roland ix
Heverly, Clement F. 186
Hickman Mills 150, 154
Higbee, John 15, 38, 150, 172–173
Higbee's Mill 38
Hirsch, Steve 186
Historical Society of Pennsylvania xiii
The Historical Society of the Cocalico Valley xii
Hockensmith, Charles D. v, vi, xiv, 1, 2, 9, 14, 17, 135–136, 141–143, 181, 182, 186, 187
Hockensmith, Susie xiv
Hogg, Gladys xi
Holleman, A. B. 191
Holloway, John 173
Holmberg, Jim xiii
Hoover, Henry 152
Hoover, Jacob 152
Hörter, Fridolin 138, 187
Howell, Charles 188
Hubble, Anna Joy Munday 188
Huff, Valentine 15
Hughes, William C. 188
hulling buckwheat 8
hulling of rice 8
Hurley, New York 37

Indian corn 18
Indian Creek 14
Inman, Gloria Kay Vandiver 184
The International Molinology Society of America xiii
Ireland, John 173
iron bands 8, 155, 165
iron wood 21
Iron Works Road 16
Irwin, John Rice x, 188
Ison, Cecil R. x, 188

Jäckle, Hans Werner 137
Jackson, Claude V. x
Jackson, Ronald V. 188
Jackson Purchase region, Kentucky 14
Jagailloux, Serge 188
Jamaica Township, New York 38
Janies, Thos. G. 177
January and Huston 156
January and Sutherland 154
Jegli, John B. 188
Jobey, George 139, 188
Johnson, Bertrand L. 188
Johnson, Christina 28
Johnson, Francis 155
Johnson, Isaac 28
Johnson, Joseph R. 41, 188
Johnson, Kevin xiv
Johnson, Martin 15, 38, 39, 41, 159, 175–176
Johnson, Robert 17
Johnson, William 39, 175
Johnston, David 156
Johnston, M. 11
Jonathan Creek 14
Jones, Amanda C. xii
Jones, Walter B. 188
Judd, William 188

Kash's Knob 15
Katz, Frank J. 188, 189
keel boots 155
Keith, John R. xiii
Kelleher, Tom 189
Keller, Allan 168
Kelly, William xi
Kennedy, D. M. 190
Kenton County, Kentucky 18
Kentucky Archaeological Survey xiv
Kentucky Commerce Cabinet xiv
Kentucky Department of Libraries & Archives x, xiii, 25, 172, 173, 175, 177
Kentucky Gazette 15, 25, 149, 150, 152–155, 157, 189
Kentucky Geological Survey xii, xiv, 9, 11
Kentucky Heritage Council ix, xiv, 3
Kentucky Historical Society x, 23, 26, 43
Kentucky Nature Preserves Commission xi
Kentucky Reporter 15, 158, 189
Kentucky River 15, 149
Kentucky Sentinel 25
kevels 137
Kidville, Kentucky 40
Killebrew, J. B. 189
Kimbrell, Johnny ix
Kingston, New York xii, 37, 42
Kinnaird, Lawarence 189
Kit Point knob 8, 84, 98, 178
Knobs Region, Kentucky 8, 15, 19, 20, 148, 160
Knox, Steve xii, 123–127
Krauss, Jewell Bell 189

La Ferté-sous-Jouarre, France 138, 164
Ladoo, Raymond B. 189
Lake Erie region 155

Index

Lancaster County, Pennslyvania xii, xiii, 135–137, 140, 163, 166
Lancaster County Historical Society xiii
Laurel Hill millstones xiii, 152, 154–156, 164
Lawrence, Vincent xii, 139
Lawrence, Wally xii, 139
Lawrenceburg, Kentucky 11
Lawson, Rowena 189, 190
lawsuits 25
Leahy, P. Patrick xiii
Lee, Jacob xiii
Lee, Rhondle ix, 3
Lee Formation 19
Leffler, Dankmar 138, 190
Lemon, J. R. 190
Letcher County xi, 3, 11, 12
leveling cross 59, 73–74, 94–95, 109, 115–117, 134–135, 145, 164
Lexington, KY 38, 150, 152, 154–155
Lexington Daily Press 25
Lexington Observer & Reporter 25, 154, 155, 190
Licking County, Ohio 164, 165
limestone 17, 18
Lincoln County, Kentucky 39, 40
Linney, W. M. 190
Little Stoner Creek 26
Lloyd, Matilda 38
Logan, Benjamin 14, 26, 148
Logan County, Kentucky 3, 11, 13
Loughridge, R. H. 190
Louisville, Kentucky x, xii, xiii, 153, 155
Lulbegrud Creek 15, 26, 30, 33, 36, 44
Luzader, David 29
Luzader, Sandra H. 29
Lynch, Michael J. xiv
Lynchburg Daily Virginian 190
Lyons, Pam x, xiii

Madison, Indiana xi, 3
Madison County, Kentucky 11, 13, 150, 154
Main Elkhorn 12
Mainz, Germany xiii
Major, J. Kenneth 138, 190
Making of America books 25
Making of America periodicals 25
Mangartz, Fritz xiii, 138, 185, 190
Manufacturer and Builder 190
Marbletown, New York 36, 37
markets 150–151
Marks, Annie L. 182
Marquet, Cynthia xiii
Marshall County, Kentucky 3, 11, 14
Mather, William S. 18, 190
Martin, Larry 190
Martin, Tom x
mauls 137
Maxwell, Hu 190
Maysville 154, 156
The Maysville Eagle 153, 156, 194
McAfee, Robert 12, 185
McCall, John M. 154
McCalley, Henry 190
M'Cauley, J. H. 190
McClain Printing Company xiii

McConathy, I. I. 150
McCorga, Radford 172
McCoum and Kennedy 12
McCreary County, Kentucky 3, 11, 14
McDowell, Robert C. 190
McGee, Marty 190
McGill, William M. 190
McGrain, John W. xi, 190
McGuire, John 33
McGuire, Kelly xi
McGuire Millstone Quarry xi, 23, 47–63, 179
McIlhaney, Calvert W. x, 190
McKechinie, Jean L. 190
McKee, Harley J. 190
McKenley, John 27
McKinley, John 32
McMurtrie, H. 190
Meadows, Larry G. ix, x, xii, 1, 2, 3, 15, 26, 30, 47, 123, 143, 153, 187, 190
Meadows, Leif ix, x, xii, 123, 124
Metcalf, Robert W. 191
meuliers 165
mica 8
Michael, Ronald L. 191
Michaux, F. A. 191
Middlewood, Esther xiii, xiv
The Mill Monitor xiii
Miller, Henry 11
Miller, Mrs. Ralph G. 191
Miller, Railsback & Miller 11, 12, 157
Miller, Raymond xiv, 9
Miller, Sarah xiv
Miller, Tamara G. xiii
mills 1, 7, 12, 13, 152
millstone cutters 164
Millstone Edge, England 134
millstone factory 164
millstone grit 165
Millstone Hill, Kentucky 14
millstone pick 165
millstone recording form 22–23, 167–169
millstone rejection 132–133, 135–136
Millstone Ridge, Kentucky 13
millstone values 12, 16, 17, 157–158
millstone size ranges 113–114
millstones 1, 7, 9, 51–60, 67–74, 78–83, 86–95, 100–102, 106–110, 112–114, 132–133, 135–136, 138–147, 165
Mineral Resources of the United States xiii, 165
Mineral Yearbook xiii, 165
Mississippi River 149
Mitchell, Steve xiv
monolithic millstones 7, 8, 165
Montgomery County, Kentucky 14, 15, 25, 26, 27, 30, 32, 33, 35, 36, 38, 39, 40, 41, 42, 43, 44, 159, 172–173, 191
Montgomery County, Virginia 136, 163
Moore, Jacob 175
Moore, Lucas 191
Moore County Grit Millstones 165
Morehead, Patsy 35
Morgan, David L. ix
Morgan's Station 26
The Morning Star 194
Morris and Egenton 153

Morris Mountain 15
Morton, Lynn Douglas 191
Morton, Mildred Napier 191
Mount Airy, Pennsylvania 140
Mount Sterling 26, 38, 39, 43, 148
Mt. Sterling Wigg 25
mountain laurel 21
Mount-Vernon, Kentucky 16
Moyland, John 149
Mullenax-McKinnie, Michelle L. xiii
Murphy, Rose xiv
Museum of Appalachia x
Muskingum County, Ohio 164–165
mustard 8
Myers, W. M. 189

Nagle, William D. 136–137, 140
Nason, F. L. 191
National Register of Historic Places 3
Neal, Robert 12
Neary, Donna M. ix
Nelson, Alfred 28
Nelson, Charlott 39
Nelson, William 28, 39
Nelson, William H. 39
New Amsterdam 37
New Harmony, Indiana 13
New Jersey 37
New Orleans 149, 155
New York xi, 7, 37, 124, 135–137, 139, 160
New York State Museum Bulletins xi, xiii, 191
New Winchester, Indiana 43
Newland, David H. 191, 192
Nicholson, Robt. 173
Niquette, Charles M. x
Norris, Tennessee x
Norris, William V. 192
North Carolina xii
North Carolina Land Company 192
North Carolina State Geologist xii, 192
North Elkhorn Creek 17
Northern red oak 20
Northumberland, England 139
Nova Scotia 155

Oak Ridge Mills 154
Observer and Reporter 150, 192
Ohio 19, 37, 156
Ogle, Alexander's Mill 153, 155
Ohio River 155
Ohio Valley Historical Archaeology xiii
Old Lexington Steam Mill 150
Old Mill News xiii, xiv
olive oil 8
O'Malley, Nancy v, xiv, 1, 2
out crops 165
Owen, David Dale 192
Owen, Katheryn 192
Owen County, Indiana 28

paint mills 8
paint staff 165
The Palladium 17, 25, 194
Palmer, William 127, 173–174
papaw 21
Paris, France 164
Paris, Kentucky 16

Paris Western Citizen 153, 155, 192
Parker, Edward W. 192
Parker, Ophelia 35
Parrish, Greenberry B. 146
Parsons, West Virginia xiii
paste 8
Patrick, Tracy R. 192
Peak District, England 134–135, 137–139
Peak millstones 165
pearling barley 8
Pembroke, Virginia x
Pennsylvania xi, xii, 7, 32, 37, 135–136, 139
The Pennsylvania Magazine of History and Biography xiii
Pennsylvanian geological period 8, 19, 20, 179
Perkins, William 149
Perry, Polly 39
persimmon 21
Peters, J. T. 192
Phalen, W. C. 192, 193
phosphate rock 8
photographs xiv
picks 137
Pigg, Anderson 27, 31, 32, 33, 39, 40, 43, 159
Pigg, Elizabeth 40
Pigg, Polly 40
Pigg, William 39
Pigg, Woodford 40
pignut hickory 21
Pilot Knob 8, 19, 20, 29, 31, 104, 165, 178–179
Pilot Knob Millstone Quarry xi, 15, 104–115, 117–121, 123, 127–131, 133
Pilot Knob State Nature Preserve 47
pine 20
Pine Mountain, Kentucky 12
pitchers 137
pitching tools 137, 165
pits 7, 49, 64–66, 77, 85–86, 98, 100, 104, 165
Pittsylvania County, Virginia 39, 40
plaster 8
plaster of Paris 8, 155, 156
plugs and feathers 136–137, 139, 165
points 124, 137, 146, 165
poison ivy 21
Pollack, David xi
Polsgrove, Tracy A. xiv
post oak 21
Powell, Everett 29
Powell, Pinkie 29
Powell, Vina 29
Powell County, Kentucky xi, 3, 4, 5, 7, 8, 9, 11, 14, 15, 19–135, 143–147, 148–151, 157–161, 163–165, 178
Pratt, Joseph H. 193
Prewitt, Anna 37
Price, Jimmie L. 187
Prichard, Jim xiii
Prince George County, Virginia 140
proof staff 165
Prussian 18
The Public Advisor 153, 194
pudding stone 18
punches 137
Pung, Olaf 190

quarry 165
quarry excavations 122–123
quartize 8
quartz 8
quartz pebbles 8, 13, 14, 15, 17, 19, 179–180
Queens, New York 38
querns 8, 165

Raccoon Burr millstones 152, 156, 165
Radley, Jeffrey 193
Rapp, Frederick 13
Rawson, Marion N. 193
red cedar 21
red elm 20
red maple 21
red mulberry 21
red oak 20, 21
Red River 19, 28, 31, 32, 44, 149
Red River Historical Society and Museum x, xi, xii, xiv, 3, 123, 124
Red River Iron Works Road 148
Red River Millstone Quarry 15, 26, 27, 28, 29, 32, 35, 36, 38, 44, 46, 150, 156, 178–179
Red River millstones xiii, 14, 15, 150, 152, 154, 155, 157–158, 159, 165
redbud 20, 21
regrinding middlings 8
Reinemund, John A. 193
Reynolds, Isom 41
Rice, C. L. 193
Rice, David 12
Ridenour, George L. 193
ridge tops 48, 50, 63–65, 85, 98, 105
Risk, Polly Anderson 41
Risk, Rachel Miller 41
Risk, William 14, 15, 26, 40, 41, 148
Robertiello, Barbara 193
Roberts, Ellwood 193
Robinson, Gary xiv
Rochester Township, New York 37
Rockcastle County, Kentucky 3, 11, 15, 16, 157
Rockwell, Peter 193
Rodes, William 158
Rogers, Diane 193
Rogers, John 154
Rogers, Lillian Berry 193
Rogers, Steve x
Rogers, William Barton 193
roller mills 9, 165
Römisch-Germanisches Zentralmuseum, Germany xiii
Rosakranas, Eleanor S. xii
roses 21
Ross, John 27, 28
Ross, John A. J. 28, 41
Rotenizer, David E. x, 193
Rotten Point knob 8, 19, 47, 48, 63, 76, 165, 178
Roundstone Creek 15, 16, 17
Rucker, James 12
runner stone 7, 16, 159, 166, 172–173
Russell, John 193
Russell County, Kentucky 13, 154
Rutledge, Clayton 31
Ryan, Thomas R. xiii
rynd 7

Safford, James M. 193
Sage Point 180
Salt River 17
Sanders, Thomas N. ix
sandstone 8, 17, 19, 166
Sanford, Samuel 193
Sass, Jon A. 193
sassafras 21
Saville, W. C. xi, 135–136, 141–143
sawbrier 21
Saxony, Germany 137
scarlet oak 21
Schauble, M. 188
Schmidt, Robert G. x
Schrader, Frank C. 193
Schulte, Herman 18
Scientific American 156, 193
Scott County, Kentucky 30
Sentinel Democrat 25
service berry 21
Shacklett family xii
Shagback hickory 20, 21
Shane, John D. 14, 26, 41, 148. 193
Shanks, David 46
shaping debris 120–121, 166
sharpening millstones 166
Shawagunk, New York 37
Shawneetown 13
Shearer, David x
shellbark hickory 20
shelling oats 8
Sherrick, Yvonne xiv
Shirley, George S. 16
Shrouse, Henry 152
Silver Creek 13
silver maple 21
Simpson County, Kentucky 13
sketch map 22
Skiba, John B. xiii
sledge hammers 136–137, 142, 144, 146, 163, 166
Sleds 148
slippery elm 21
Smith, Robert C., II xi
Smith, Sara A. 41
Smith, William 41
Smith, William M. 41
smut machines 156
snuff mills 8
Society for the Preservation of Old Mills xii, xiii, xiv
soils 19, 20
Somer, Cornelius 33
Soper, H. L. 153, 154
Sopko, Joseph 193
sourwood 21
South Union, Kentucky 13, 149
southern blackhaw 21
southern red oak 21
Spafford, Horatio Gates 194
Spanager, Nicholas 18
Spencer-Morton Preserve 20
spicebush 21
spices 8
spindle 7, 153, 155
split peas 8
spoons 137
Spring, Steve xiii
Spry, Cornelius 41–42, 159, 176–177
squares 137, 143, 146–147

Index

Stansefer, Gabriel 12
Stanton, Kentucky 49
State Historic Preservation Officers xi
Station Camp Creek 13
Stedman, E. 155
Stedman, H. 155
Sterrett, Douglas B. 194
Stewart, Ann 42
Stewart, Eli 42
Stewart, Elijah 42
Stewart, J. E. 28
Stewart, James 42
Stewart, Sydney 42
Stoddard, B. H. 183
Stokes, Julia S. xi
Stone, Ralph W. 193
stone cutters 45–46, 166
Stone Quarry Road 148
Stoner, Jacob 194
Stoner Creek 153
Storck, John 194
Strawhacker, William 194
striking hammers 137, 166
sugar maple 20
Sullivan, George M. 194
Summers, Cornelius 15, 27, 42, 43, 159, 177–178
Summers, Elizabeth 27
Summers, Rev. C. 43
Suesserott, J. L. 190
Surface, Robert Houston xi, 135–136, 141–143
Sussenbach, Tom x, 14, 17, 194
Sweet, Palmer C. xii
sweet gum 21
sweet pignut 21
Swift, Michael 194
Swisher, Jacob 194
Symposium on Ohio Valley Urban and Historic Archaeology xiii

T. L. C. Genealogy 194
Talbert, Charles G. 194
talc 8
tanbark mills 8
Tanner, Joel 27, 173
Tanner, John 17
Tates Creek Road 150, 155
Taylor, Samuel 15
Teagarden, W. H. 155
Teagarden Mill 150
Teague, Walter D. 194
Teeples, Gary R. 188
Tennessee 13
Tennessee Historical Commission x
Tidewater Atlantic Research x
Tingle, Tim xiii
Todd, Levi 194
Toler, Trigger xi, 194
Toler Millstone Quarry xi, 23, 30, 76–84, 112–115, 117, 119–123, 127–133
Tomlinson, Tom D. 194
tool marks 59, 63, 74–76, 83–84, 95, 97, 102, 109–110, 127–132
tools xii, 136–137, 166
transportation 148–149
Treadway, Abigail 28, 33
Treadway, Catherine DeWitt 43
Treadway, Christine 28
Treadway, Joel 32

Treadway, John 28, 32, 43
Treadway, John, Jr. 43
Treadway, John, Sr. 43
Treadway, John D. 43
Treadway, Margaret 32, 44
Treadway, Margerie 32, 33
Treadway, Moses 26, 27, 32, 33, 43–44, 159, 173–174
Treadway, Moses H. 44
Treadway, Moses X. 44
Treadway, Nancy 28
Treadway, Peter 28, 44, 46, 159
Treadway, Peter D. 28
Treadway, Polly M. 43
Treadway, Thomas 28
Treadway, William 28, 44
Treadway, William E. 194
Tribble, Malinda 38
Truax, J. W. 194
tubmill 154
Tucker, D. Gordon 134, 138, 194, 195
Tucker, Gertrude E. 182, 183, 184
tuliptree 20, 21
Tuomey, M. 195
Turkey Hill Millstone Quarry xii, 135, 166

Ulster County, New York, millstone quarries xii, 139
Ulster County, New York xii, 36, 37, 42, 135–136, 139, 164
United States 3, 5, 7, 9, 12, 24, 135, 138
United States Bureau of Mines and Minerals 166
United States Forest Service, Winchester, Kentucky x
United States Geological Survey xiii, 134
United States Secretary of State 195
University of Kentucky x, xiv, 2, 24

Vandike and Keller 12
Verhoeff, Mary 195
Versailles, Kentucky 12
Vicar, D. M. 154
Vinton County, Ohio 165
Virginia xi, xii, 7, 32, 35, 36, 39, 40, 41, 124, 135–137, 139–140, 143, 156, 160
Virginia pine 21

wagon 14, 15, 26, 148
Walden and Ward 156
Wales 134–135
Wallace, Susan 31
Wallcut, Thomas 195
walnut 20
Ward, Owen H. 134, 138, 195
Ware, Achilles 29, 45
Ware, Dillard 45
Ware, Dillard P. 28, 44
Ware, Dora 45
Ware, John 29
Ware, Killis 45
Ware, Martha C. 28, 45
Ware, Pattie A. 45
Ware, Thomas B. 28, 45
Ware Millstone Quarry x, xi, xii, 29, 84–98, 112–115, 117–121, 123, 125–131, 133
Warren County, Kentucky 13
Waruch, Lewis xii, 139
Warwasing, New York 37
Washington, North Carolina x
Watson, Thomas 195
Watson, William 154
Watson's Mill 154
Waynesboro, Pennsylvania xii
Weaks, Mabel Clare 195
Weaver, Valerie 195
Webb, Wayne ix, x, xii, 123–127
Webb, William S. 195
Websites 3, 25
wedges 124–125, 137, 165
wedges and feathers 136, 144–145
wedges and shims 136
wedges and slips 136, 141
Weir, G. W. 193
Welch millstones 134
Wesler, Kit xiii
West, Donna 45
West, Harrison 28, 45
West, Levina L. 28
West, Mary E. 28
West Virginia 38
wheat 8, 17, 152, 156
whiskey 15, 172–173
white ash 20, 21
white basswood 21
white hickory 21
white oak 20, 21
Whitley County, Kentucky 3, 11, 17
Whyte, Samuel G. 13
Wieck, Dorothy L. 195
wild blackcherry 21
wild hydrangea 21
Wilderness Road 156
Wilkerson, Joseph 15, 26, 173
Wilkie, Aitken 195
Wilkinson, James 149
Williams, Albert, Jr. 195
Williams & Company 195
willow 21
Wilson, John 16, 27
Wilson, John, Jr. 33
Winchester, James 13, 154
Winchester, Kentucky x, xiii, 26, 32, 35, 37, 38, 41, 45, 148, 155, 158, 176
Winchester Democrat 25
Winsor, Joseph H. 181
Winter, Alisha I., Jr. 154
Wolf, Carol L. 185
Wonn, Mildred 195
wooden wedges 137
Woodford County, Kentucky 3, 11, 12, 17
Woodstock, New York 42
Worsham, Gibson 195
Wright, Lucy 42

yellow pine 21
yellow poplar 21
Young, Wingate 28, 29
Young, O. 173
Youngs, Benjamin S. 13

Zerfass, Samuel G. 195

www.ingramcontent.com/pod-product-compliance
Lightning Source LLC
Chambersburg PA
CBHW081556300426
44116CB00015B/2902